Managing
Natural
Resources
with GIS

Laura Lang

ESRI
    Managing Natural Resources with GIS
    ISBN 1-879102-53-6

First printing June 1998. Second printing May 2000. Third printing August 2001. Fourth printing February 2003.

Printed in the United States of America.

Published by ESRI, 380 New York Street, Redlands, California 92373-8100.

Books from ESRI Press are available to resellers worldwide through Independent Publishers Group (IPG). For information on volume discounts, or to place an order, call IPG at 1-800-888-4741 in the United States, or at 312-337-0747 outside the United States.

Contents

Preface

Between the end of this century and the middle of the next, the human population will double to 12 billion. This means that we will increasingly have to squeeze this little planet of ours for more food, water, and fuel. As a result, managing earth's limited natural resources has emerged as perhaps the most crucial problem now faced by humanity.

Fortunately, technologies are now becoming widely available that may allow us to feed and power the growing population without destroying the very environment that sustains us in the process. And not a moment too soon.

With this technology we are starting to measure virtually everything on earth and how these things move and change over time. Fed into spatial databases called geographic information systems (GIS), these measurements help us understand what's happening all around us and how to improve it.

Managing Natural Resources with GIS shows how a dozen different organizations around the world use GIS for everything from precision farming to forest management. It even shows how companies that are resource users (rather than managers) are using GIS to protect endangered species and restore sensitive environments while still turning a profit.

It is my hope that all of the competing interests around the globe will eventually use GIS to find some common ground.

Jack Dangermond
Founder and president
ESRI

Acknowledgments

Last year, we asked our offices worldwide to recommend natural resource projects in their regions that could be profiled in a new book. Our interests ranged from fisheries to forests, from urban planning to parklands.

Over several weeks, we received an outpouring of information about natural resource projects in Europe, Asia, South America, Africa, North America, and even Antarctica. I would especially like to thank Bob Ruschman, Jesse Theodore, Matt Artz, Karen Hurlbut, Charles Convis, Lee Ross, Dale Loberger, Kevin Daugherty, Bonnie Mananua, Alfonso Rubio, Frank Holsmuller, David Maguire, John Steffanson, Sara Moola, and Lee Ross for the projects they suggested and the contacts they provided.

Although we could include only 12 user case studies in this book, the many we heard about helped us understand the enormous challenges faced by governments, conservation groups, schools, and individuals who are using GIS software to help preserve and wisely use earth's resources.

The users whose projects we've included spent many hours explaining their local resource issues, reviewing drafts, and supplying illustrations. They are acknowledged by name at the end of each chapter. Each was essential to the success of this book.

Michael Karman and Tim Ormsby edited the book. Michael Hyatt designed the book, laid out and produced the pages, and did the copyediting and proofreading. Gina Davidson designed (and redesigned) the cover as the book went through various titles and cover-illustration concepts. Cliff Crabbe oversaw print production.

This book was made possible by Bill Miller, Judy Boyd, and Christian Harder, who gave me the opportunity to work on a book for the first time. And, finally, a special thanks to Jack Dangermond, whose dedication to forwarding the role of GIS in natural resource endeavors got the whole project rolling.

Chapter 1

Managing natural resources

At a time when human population growth is taxing the earth's abundance as never before, natural resource managers are discovering the power of geographic information systems (GIS) to help them make the crucial decisions they face every day. Once an expensive technology favored by research scientists, GIS emerged in the 1990s as the tool of choice in local, state, and national resource agencies around the globe. GIS is helping the development and conservation communities find common ground by providing a framework for the analysis and discussion of resource management issues.

The role of GIS

Biologists, botanists, planners, petroleum engineers, foresters, and corporate executives are increasingly relying on GIS to help them make critical decisions. By putting their spatial data in an integrated system where it can be organized, analyzed, and mapped, they find patterns and relationships that were previously unrecognized. This in turn gives them a deeper understanding of the issues they face, and lets them bring more information and less conjecture to the problem-solving process.

Whether they're restoring habitat, planting a vineyard, searching for oil, fighting wildfires, or measuring endangered species populations, these "spatially literate" users have learned to unleash the power of GIS for managing natural resources.

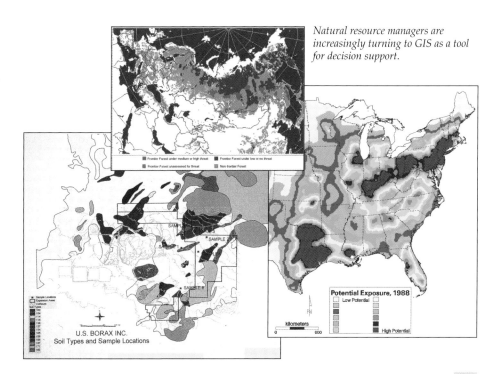

Natural resource managers are increasingly turning to GIS as a tool for decision support.

The goal of sustainability

There's no simple answer to the environmental problems confronting the world. People need to live, to eat, to improve the quality of their lives, and this means that resources will be consumed. Too often, however, simple ignorance (lack of information) leads to practices that use resources in ways that can't be sustained over time. And in a very real sense, our ability to sustain resources may determine our fate on the planet.

But there is increasing optimism in many circles that the application of information technologies to natural resource management will help us feed and house 6 billion people next year, and perhaps 10 billion by 2020, without ravaging the environment.

Two trends in particular—the proliferation of measuring devices and advances in geographic information systems—offer the best hope yet for comprehending and correcting the damage being done to our planet.

We live in an age when just about everything that moves or changes over time is being measured. Remote sensing technologies are creating data faster and in greater volumes than ever before, and all of this information is geographically referenced.

Fortunately, we also live in an age when the computing power and information management tools are in place to allow people to use this data productively. These geospatial tools make it possible to focus our attention on the large problems, to pinpoint the small ones, and to anticipate those that are waiting in the wings. Natural resource managers are increasingly turning to GIS as the crucible in which this data can be processed and from which solutions can be drawn.

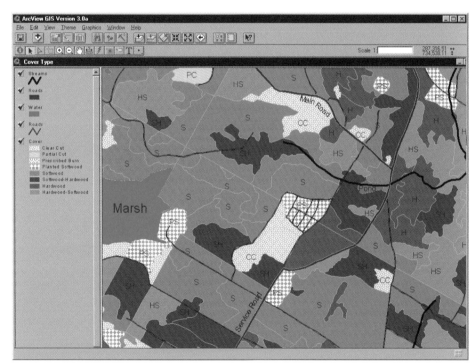

Foresters at the University of New Brunswick, Canada, use GIS to manage the woodlot shown above. By carefully selecting when and where to cut trees, foresters ensure that the woodlot continues to produce timber while remaining a balanced ecosystem.

Real-world examples

The 12 case studies in this book cover a variety of environmental applications for GIS. In some of them, the systems are run by nonprofit organizations monitoring a particular ecological threat, like deforestation. In others, GIS is used by public agencies to help restore the natural habitat of endangered species or to reclaim areas of polluted industrial land. In still others, heavy industries themselves use GIS to help them extract essential resources, like oil and minerals, with minimal impact to the environment.

It wasn't possible, of course, to address every issue of environmental resource management, but an effort has been made to be representative. The chances are that even if your particular field isn't covered, you'll see familiar concerns and find useful problem-solving strategies.

If you already know the basics of GIS, you may want to skip directly to the case studies. If not, it may be worth a few moments of your time to read the introductory material that follows.

WORLD RESOURCES INSTITUTE

DCMP
DELAWARE COASTAL
MANAGEMENT PROGRAM

Portland Metro

City of New York

What is GIS?

What is a geographic information system? No one has yet found a definition that satisfies everyone. Essentially, though, a GIS is any of various software applications, running on PCs or workstations, that stores, analyzes, and displays multiple layers of geographic information. In simplest terms, a GIS can be thought of as a spatial database.

What does that mean exactly? First, it means that discrete geographic locations on the earth's surface can be stored in computer files as sets of mathematical coordinates. This makes it possible to draw a map on a computer: a map of the world, a map of the Amazon basin, a map of your neighborhood.

Second, it means that different map files, or *layers*, of spatial information with common geography can be displayed simultaneously and analyzed with reference to one another. In a map of an agricultural area, one layer might represent the boundaries of the land, another the soil types found there, another the local streams, and still another the changes in elevation.

The analytical power of a GIS lets you query the system to extract information from multiple layers. For instance, to find the most suitable place to plant a crop, you might identify locations with a particular soil type, lying at a certain elevation, and receiving a specified amount of rainfall.

Third, it means that any quantitative information that can be linked to geography can be used in a GIS. Not only can you represent, for example, the locations of toxic waste sites in a given region as points on a map, but you can symbolize these points (draw them in different colors, sizes, and shapes) according to any information you have about them, such as the type of waste they store.

Fourth, it means that geographic features and phenomena can be modeled from sample data. A typical example is that of a digital terrain model. Sample elevation data is gathered at various points and the GIS uses this input to create a continuous elevation surface, or, in other words, to build a model of the landscape. Similarly, models of processes, such as the spread of oil slicks or wildfires, can be simulated from sample data and from assumptions about the movements of forces like winds and tides.

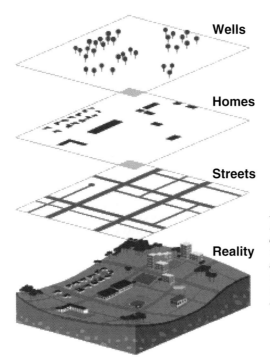

Wells

Homes

Streets

Reality

A GIS stores information about the world as layers of spatial features (wells, buildings, streets, and so on). These separate layers can be interrelated on the basis of shared geography. This simple but powerful concept has proven invaluable for solving many problems: from tracking endangered species to modeling circulation of the atmosphere.

Why geography matters

The study of environmental resource management makes it clear that what is taken with one hand is very often given by the other, whether we know it at the time or not. The city of Los Angeles flourishes on water diverted from the Owens Valley. The result is that Owens Lake dries up, leaving a bed of salts and toxins that become a major source of airborne pollution. Economic prosperity builds skyscrapers in Chicago, but builds them in the flight paths of migrating birds, who crash by the hundreds of thousands into reflected skies. Rain forests are cut down for timber and to clear arable land, and the global implications—for plants and animals, rivers and oceans, and the earth's atmosphere—boggle the informed mind.

GIS is an ideal system for analyzing the impact of development and consumption on natural resources, because geography is the playing field, literally, on which these dynamics unfold. Information maintained by different organizations, often for very different purposes, can be integrated and analyzed to visualize relationships, find explanations, and develop solutions to pressing problems.

Consider a wildlife manager concerned about the decline of a fish species in a stream network. She studies the fish's local habitat and finds nothing apparently wrong. With GIS, she can bring more information to bear on the problem.

A map of the larger region shows her an important stream that links her fish population to another with which it breeds. Taking samples of the stream water, she finds that the fish may be avoiding it because the temperature is too high. Has something changed?

Again, GIS can help. Land use data from a city planning department reveals that forested land along the stream was cleared for housing developments. The trees that once cooled the water with their shade are no longer there, and a plausible explanation for the problem has been found.

What about a solution? Perhaps measures can be taken to cool the stream. Adding large amounts of woody debris to the stream itself provides shade that lowers the water temperature. Or perhaps other connecting streams can be rendered more suitable to the fish. In any case, GIS has been crucial in establishing an unexpected relationship between the building of houses and the welfare of a wildlife population. This knowledge can be used to improve resource management in the future.

Education and spatial literacy

At least as important as good resource management is environmental education. And in dealing with issues so fundamentally rooted in space, the language of geography is what's spoken. GIS increases spatial literacy—the understanding of how issues of place affect the decisions we make—by putting the tools of geography in the hands of many, and making them easy to use and understand.

As people come to appreciate the interconnectedness of environmental problems, the chances for cooperation among nations, international organizations, and interest groups of all kinds improves. And as spatial literacy improves, this goodwill can be translated into increasingly effective action.

Ultimately, we need to make sure that we raise a new generation of environmentally responsible citizens. One of the best ways to do this is to bring GIS into the classroom. As this book's case study in education shows, high school students with the right training and materials are as capable as adults of carrying out projects that make a difference.

Chapter 2

•••• Oil and gas exploration

Oil and gas companies that explore and drill all over the world must meet the environmental requirements of their host countries, to interfere with local communities and fragile resources as little as possible. Using GIS and new drilling technologies, they can place surface collection and processing facilities miles away from sensitive areas and still reach valuable underground reserves.

In this chapter, you'll see how Chevron, a major international oil company, uses ArcView® GIS software to locate oil in the Niger Delta and work with the Nigerian government on long-term agreements for oil extraction and resource protection.

Drilling in sensitive areas

Chevron Nigeria Ltd. (CNL), a partnership between the Nigerian National Petroleum Company (NNPC) and the California-based Chevron Corporation, produces about one-fifth of Nigeria's oil. Since 1961, CNL has grown steadily, helping to strengthen the petroleum industry of western Africa, one of the world's most important oil-producing areas.

In 1995, CNL began to implement GIS technology. GIS has already become integral to the partnership's operations, allowing it to drill and process oil with the least possible disruption to Nigeria's plants and animals.

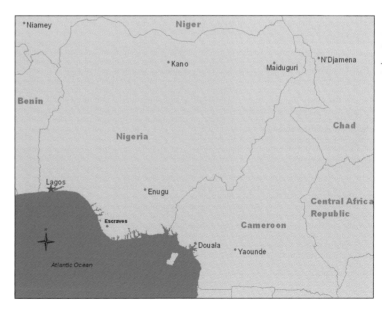

CNL produces about 400,000 barrels of oil a day, much of it from wells along the Niger Delta.

The concerns

The Nigerian government is naturally concerned about the environmental hazards of oil production—drained wetlands, polluted rivers and streams, vanishing fish and wildlife. Consequently, oil companies working in the region must meet strict environmental guidelines, and work with the government when exploring for oil reserves, building roads, dredging canals, or placing facilities near human or wildlife habitats.

Chevron's scientists imported satellite imagery and aerial survey data into their ArcView GIS system to create a base map of the region. They verified and corrected the positions of fixed features like oil wells and roads with global positioning system (GPS) receivers. Other data, such as the location of wetlands, endangered species, and human populations, was then added to the digital maps.

Employees and managers are linked to the system, using it for a range of applications, from oil exploration projects to planning oil spill response systems.

The Niger Delta is one of the world's largest wetlands, with the most extensive mangrove forest in Africa. With GIS, Chevron employees can locate and drill for oil with minimal impact on sensitive ecosystems—on land, in the marsh, and in the ocean.

GIS for exploration and production

When exploration geoscientists at Chevron look for oil, they use ArcView GIS software to plan seismic surveys. Before they perform the surveys, they overlay information about roads, topology, previous seismic data for the area, and land ownership (to get the necessary permits).

A seismic survey is conducted with the aid of energy sources, such as Vibroseis trucks or small explosive charges, that send shock waves into the earth. Depending on the type of rock they hit, the waves bounce back differently, enabling geoscientists to create a three-dimensional map of the subsurface. The 3-D images, produced with specialized software, are like X rays of a slice of earth. Experts can view these images and make decisions about where pockets of oil might be located.

When seismic surveys indicate oil, Chevron's production engineers overlay GIS data about roads, people, endangered species, wetlands, and other features to find the best locations for wells, collection facilities, storage tanks, roads, and pipelines. If an area above a reservoir is sensitive, the company can drill from a lower-impact area, even miles away, using directional drilling technologies that allow the well bore to be angled in.

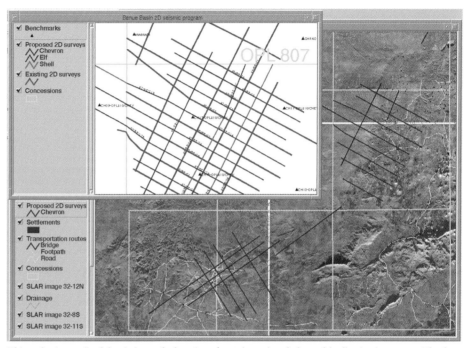

This radar image (backdrop) shows the location of past (green) and planned (red) seismic surveys. The data in the GIS helps the engineers decide where to collect seismic data. The many rivers at the bottom left, for instance, would interfere with the seismic signal. Such local impediments could previously be discovered only after lengthy, often fruitless, field trips.

Monitoring water quality

When oil is produced from underground, it contains natural gas, some contaminants (like sulfur), and water. The liquid is treated to remove impurities, and the recovered water is sometimes returned underground. Monitoring the discharge water coming from the well is necessary to determine what kind of treatment will make it safe for disposal. If contaminated water were returned underground, it could eat away at rock formations and clog the well.

Chemical analyses of water quality are kept in spreadsheets loaded in the GIS. Whenever there is an unexpected increase in any impurity or contaminant, the data is reviewed by the engineers, who may follow up with an on-site investigation.

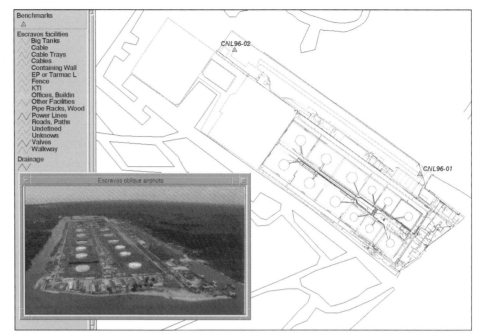

The Escravos tank farm is a Chevron storage facility where oil is kept prior to shipping. Employees are able to use the GIS to access information about Escravos and the surrounding areas, minimizing unnecessary travel to the site.

Mapping sensitive zones

Whenever engineers construct a platform for a drilling rig or dredge a canal, they use the GIS to make maps of sensitive underwater areas, like shellfish breeding grounds.

They use ArcView GIS software to overlay the locations of rigs, water depth, and sensitive zones. These maps help them plan where to route vessels and where to place platform footings and pipelines.

The GIS is also used if oil is spilled. Spill data, maintained by the Nigerian government and oil companies operating in the country, is brought into the ArcView GIS system and coded by factors like spill amount, time of day, and person or work crew involved. This helps Chevron's engineers search for common themes—like a single crew involved in many incidents. When spills happen, the GIS can be used to quickly identify priority shorelines and habitat that should be protected.

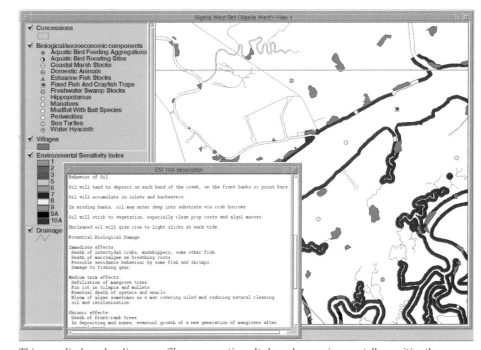

This map displays shorelines near Chevron operations. It shows how environmentally sensitive they are, and marks the locations of plant and animal habitats. In the event of a spill, crews can move quickly to protect resources.

GIS in a broader context

Chevron's public relations group uses the GIS to research communities near their operations and to select hospitals, schools, or other institutions to sponsor. Their activities, which range from donating books to building schools, often include map displays about the company and its projects to improve the environment.

In the future, CNL will work with other oil companies in the area to produce a database meeting the requirements of the Nigerian government for information about environmentally sensitive areas in the Niger Delta.

Under the auspices of the Oil Producers Trade Section, a consortium of all major oil companies operating in Nigeria, this project will be the first of its kind and will help identify and preserve the region's forests, wetlands, and other resources.

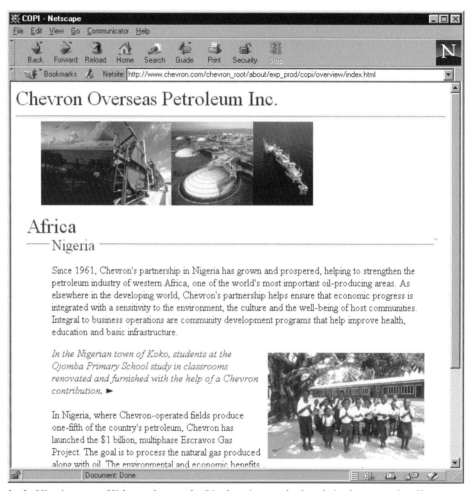

In the Nigerian town of Koko, students at the Ojomba primary school study in classrooms that Chevron helped renovate and furnish. Chevron posts its environmental programs on its Web site (www.chevron.com).

Hardware

Sun™ E-450 server

Windows NT® clients

Software

ArcView GIS for UNIX

Exceed for Windows

Data

Satellite imagery

Aerial photography

Political boundaries

Population centers

Roads

Topography

Hydrography

Vegetation

Protected areas

Endangered species

Geological maps

Geophysical maps

Seismic surveys

Well locations

Pipelines

Facilities

Acknowledgments

Thanks to William Wally of Chevron Information Technology Company, Scott Hills of Chevron Nigeria Limited, and Brad Dean of Chevron Petroleum Technology Company.

Chapter 3

Agriculture

Farming operations of all sizes use GIS to select the right growing regions for their crops and to design fields in ways that improve crop yield while saving money on fertilizers and water. Good design also reduces the potential of chemicals entering nearby streams and rivers and killing wildlife, fish, and plants. "Green" growing, as these practices are called, is applauded by consumers, suppliers, and others interested in promoting environmentally friendly business practices.

In this chapter, you'll see how an international winery used ArcView GIS software to design a vineyard in California that minimized reliance on water and artificial nutrients, while preserving the land's scenery and soils.

Finding the right region

Southcorp Wines is Australia's largest vineyard and winery company and exports many of its award-winning wines to the United States.

The company uses GIS and global positioning system (GPS) technologies to find locations with the right soil and climate for its Cabernet, Merlot, and Syrah wine grapes, and to design vineyards around the trees, hills, and streams already there.

Recently, Southcorp decided to establish a vineyard in the United States. It settled on California's central coast as the region most suitable for its grapes and, with the help of local GIS consultants, began the process of searching for the right spot.

GIS helps Southcorp Wines plan vineyards that will produce healthy crops while preserving the land's beauty and soils.

Evaluating sites for growing wine grapes

Starting in 1997, one of Southcorp's viti-culturists studied several sites ranging from 500 to 1,000 acres, all near the city of Paso Robles.

Southcorp then hired California AgQuest Consulting, Inc., an agricultural consulting firm from Fresno, California, to evaluate the physical and chemical makeup of the soil at each site. With the ultimate quality of the grapes directly linked to the environmental conditions, Southcorp was taking no chances, especially since it costs more than $15,000 per acre to develop vineyards.

One parcel, the 607-acre Creston Ranch (called the Creston 600), met the grower's criteria for soil and climate. The rolling hills of this former cattle ranch are covered with oak trees, and several natural streams flow through its valleys, which Southcorp wanted to preserve as much as possible.

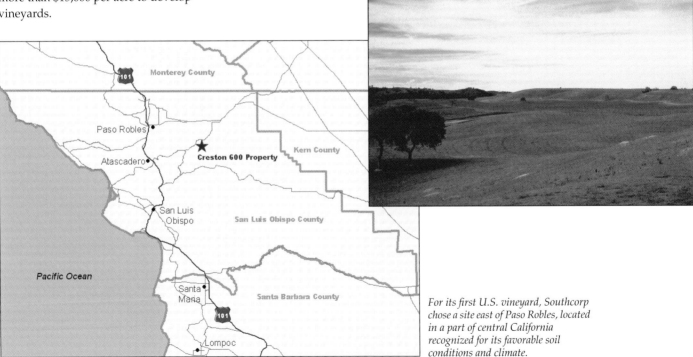

For its first U.S. vineyard, Southcorp chose a site east of Paso Robles, located in a part of central California recognized for its favorable soil conditions and climate.

Creating the property base map

AgQuest and a subcontractor, VESTRA Resources, Inc., of Redding, California, first created a GIS database of the property by combining existing data sets (roads, property lines, topography) with newly acquired data (aerial photographs, soils information, hydrology). When organized in an ArcView GIS system, the data could be overlaid into a series of maps useful for planning the vineyard.

The most important data, and the costliest to acquire, was the soil samples. Using a backhoe, AgQuest dug several hundred pits (seen at right) distributed evenly across the property. The location of each pit was recorded by VESTRA with a GPS receiver and linked to the results of chemical tests performed on the soil.

The resulting integrated GIS base map served as the foundation for a series of maps describing the ranch's soils, topography, and hydrology.

This map shows the location of all pits dug to gather soil sample data. The locations were captured with GPS receivers.

Collecting on-site data

Wine grapes are adaptable to a wide range of soil textures, from sandy loams to loams and clay loams. Each of these soil types retains water differently and therefore affects the amount of water available to the plants. From a management standpoint, it's desirable to have soils in vineyard blocks that have uniform water-holding capacities and can be irrigated for the same length of time.

Soil samples were taken at the surface and underground at 220 places throughout the ranch. These samples were evaluated to produce a profile of the physical and chemical characteristics (pH, salinity, toxicity, and nutrient levels) of the soil at each location. The data was then plotted onto maps in the form of bar charts corresponding to each sample location.

These individual charts paint a scientific picture of soil conditions from 0 to 6 feet deep. Each chart shows the soil texture, the soil consolidation, and the carbonate reaction—all crucial pieces of information.

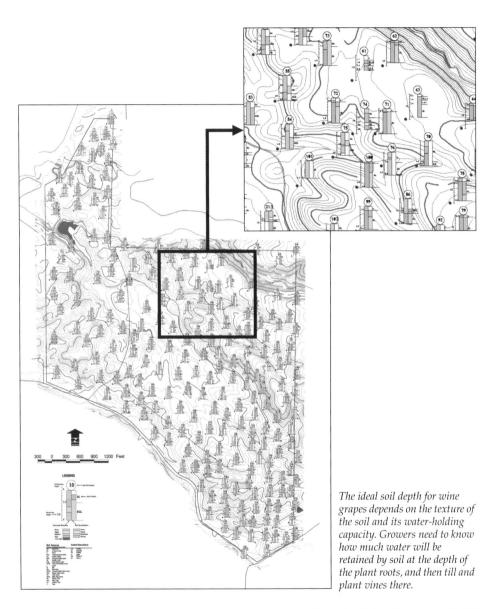

The ideal soil depth for wine grapes depends on the texture of the soil and its water-holding capacity. Growers need to know how much water will be retained by soil at the depth of the plant roots, and then till and plant vines there.

Analyzing the chemistry

Wine grapes can be grown almost any-where in central California from the standpoint of climate and soil types. How-ever, they are quite sensitive to soil salin-ity and high levels of sodium, chloride, and boron. The grapes also have difficulty with very high or low pH levels. Maps were created by the analysts to show how these chemical levels varied at the site. From these maps, the vineyard manager could decide where and in what quanti-ties to add soil nutrients, and how deeply to till the soil. They could, for instance, till the loam soil in one area at 4 feet and the sandy loam at 5.

Southcorp wanted to avoid deep tillage, down to 6 or 7 feet, which would disrupt the natural soil. Instead, they wanted options for shallow tillage so that vine roots would extend only 3 to 4 feet into the soil. At this level, it's also easier to control the water and fertilizers absorbed by the vines. The idea is to limit the plant's growth carefully, and thus allow its fruit to develop more fully.

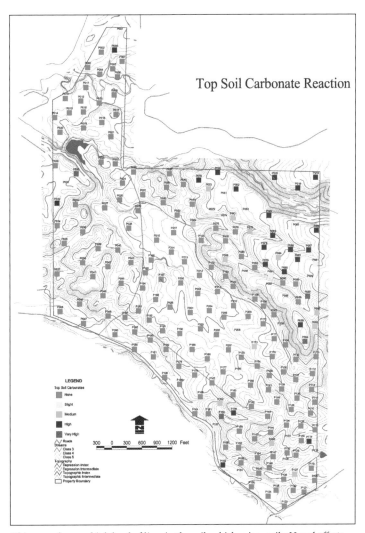

This map shows a high level of lime in the soil, which raises soil pH and affects vine growth.

Water-holding capacity

Proper irrigation has always been one of the most crucial factors in successful viticulture. Not enough water results in suboptimal crop yields, while too much water produces lesser-quality wines. The consultants developed another map with ArcView GIS software that showed how much moisture would be retained by the soils after a 3-foot-deep tillage.

Based on the measurements taken at each pit, AgQuest created the map at the right showing the water-holding capacity of the soil in each area. This data would be of utmost importance in determining where to place irrigation pipes, how far to space rows, and ultimately how much water to use once the vines were planted.

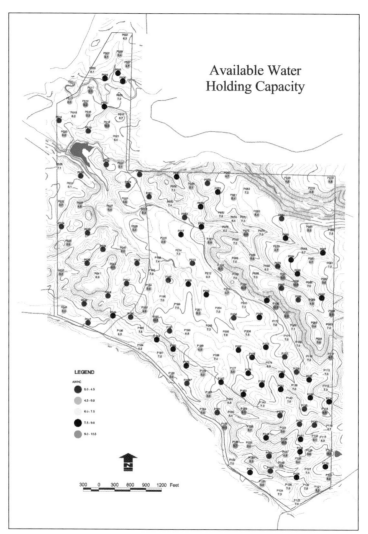

Topography and soil types help the vineyard manager plan block layouts with soils of similar water-holding capacities. This lets the grower water areas for the same length of time while preventing soil runoff from overwatering.

Planning vineyard blocks with GIS

The vineyard manager and the consultants next used ArcView GIS software to group the property into blocks according to how much moisture the soils could be expected to retain.

They also added locations of the ranch's natural scenery, such as the large oak trees and several creeks and ponds. The cluster of trees seen at the map's bottom was preserved, along with several trees growing inside vineyard areas.

Vineyard blocks would be irrigated with pipe networks covering 10 to 12 acres to minimize development, irrigation, and fertilizer costs. No riparian zones would be disturbed, nor would vine rows cross natural drainage areas, where the loss of natural ground cover caused by tilling would lead to erosion during heavy rains.

The map at the right shows planned land uses and vineyard block layouts. Four hundred and six acres were set aside for vineyards, 74 for corridors, 97 for open space, 26 for riparian zones, and 4 for housing and shops.

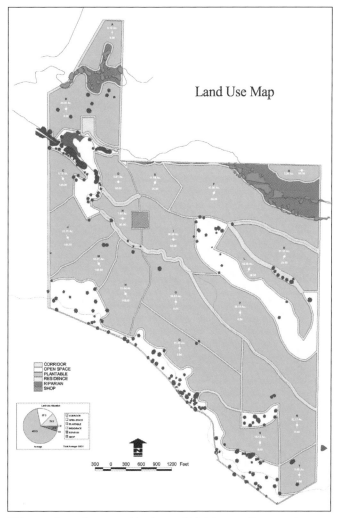

This map shows the proposed land use, which integrates the grower's environmental and business concerns.

Creating a working plan

Once the vineyard blocks were approved, a final map was created that showed the blocks overlaid with the topology of the ranch. This would allow the grower to determine the best placement for the individual rows of grapes. Like everything else about the design, the row placement was carefully thought out. First of all, the rows needed to be able to accommodate mechanical harvesting equipment. They also needed to match the natural contours of the land for irrigation purposes. And finally, they needed to be angled to gain the most exposure from the sun.

With planning complete, the GIS data can be used to estimate the amount of irrigation equipment and the number of vines, stakes, and other supplies needed to develop the vineyard. The GIS will become an information management tool and data repository, so the company's knowledge of the vineyard continues to grow along with the grapes.

Syrah, Merlot, and Cabernet were to be planted on about half of the farmable acreage starting in the summer of 1998. The first crops are expected by 2001.

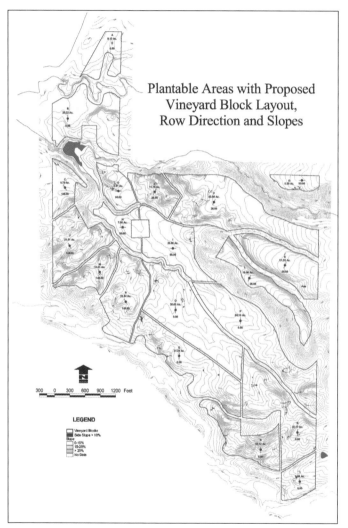

The vineyard rows are designed to accommodate mechanical harvesting and maximize sun exposure to the grapes.

Hardware

Pentium®-class PCs

Software

ArcView GIS

Data

Field-collected GPS data

Soils maps (U.S. Geological Survey and Soil Conservation Service)

Aerial photography

Acknowledgments

Thanks to Ron Brase of California AgQuest Consulting, Inc., Jamie Carothers and Don Gordon of VESTRA Resources, Inc., and Aden Brock of Southcorp Wines Estate LLC.

Chapter 4

●●●●● Deforestation

State and local resource agencies focus attention on a defined geographic region, but other groups look at the big picture of worldwide environmental change. With GIS data and software, these groups can track changes and work alongside government agencies, private companies, and the public to slow or reverse environmental degradation and promote alternative products and livelihoods.

In this chapter, you'll see how the World Resources Institute uses GIS data and maps to present the effects of deforestation to people and governments around the world.

Identifying resources at risk

The World Resources Institute (WRI), a Washington, D.C.-based policy research center, has used GIS data and software since 1994 to compile global change information and present it to researchers, academics, and the public.

The group focuses its efforts on the most dangerous worldwide environmental trends—loss of forests, ocean pollution, coastline erosion, coral reef damage—and works with governments in both industrialized and developing nations to promote policies for sustainable use of resources.

Of all its projects, perhaps none has attracted as much attention as the institute's efforts to highlight the rapid deforestation of the planet.

Worldwide, about 16 million hectares of forest are cut or burned annually, destroying habitats, threatening biodiversity, and making the land susceptible to floods and drought.

Global loss of frontier forests

WRI's Forest Frontiers Initiative studies frontier forests, defined as stands of trees in large, ecologically intact areas, undisturbed by human activities. Over the past 8,000 years, about half of the world's forests, or roughly 3 billion hectares, has been burned, cleared, or logged. These activities leave the land unable to support biodiversity, prevent erosion, or sustain indigenous people. The forests that regrow are called secondary, or fragmented, forests, and their very existence sometimes poses a threat to remaining frontier forests. In the United States, for example, the nests of songbirds in small forest patches are under attack by cowbirds, bluejays, raccoons, and other animals that thrive along forest edges.

Today, just one-fifth of the world's original forest cover remains in large undisturbed tracts of frontier forest.

To help make the scientific community, conservation groups, governments, and the public more aware of how quickly frontier forests are disappearing, a group of institute researchers used ARC/INFO® and ArcView GIS software to analyze where frontier forests still exist, and to identify the threats to their continued survival.

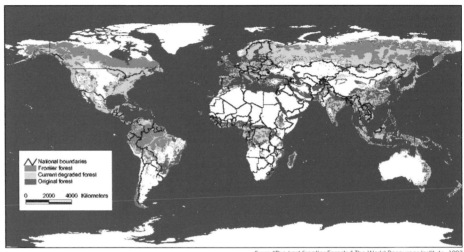

From "The Last Frontier Forests," The World Resources Institute, 1997

WRI's studies have found that just 22 percent of the earth's original forest remains in large, relatively intact stands of frontier forest. On the map, dark brown represents original frontier forest that has been lost, dark green depicts where forests remain intact, and light brown shows fragmented or secondary forests.
The widely dispersed pockets of frontier forest are mainly found in currently inaccessible regions, which has helped save them from human exploitation.

Creating worldwide resource maps

Because no worldwide digital map of frontier forests existed, the WRI scientists started with a global forest GIS data set developed by the World Conservation Monitoring Centre in Cambridge, England. They next overlaid wilderness area data from the Sierra Club. The idea was that if a tract of land was both forested and a wilderness area (not located near urban developments or major roads), then it potentially qualified as frontier forest.

Hard-copy maps of these candidate areas were sent to more than 90 forestry experts around the world for review. Areas that were found not to qualify were deleted. Other areas had their boundaries altered. The experts also answered questionnaires in which they evaluated the threats to frontier forests from such activities as logging, mining, and agricultural clearing, and rated these threats as high, medium, or low.

Edits to the frontier forest boundaries were reconciled and incorporated into the GIS database. The survey results were used to classify the forests according to risk. The GIS analysts produced a series of regional frontier forest maps, many of which can be found at the WRI Web site (www.wri.org).

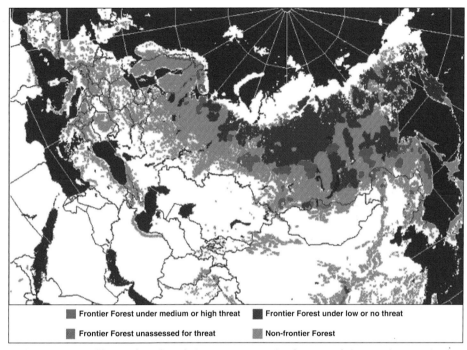

■ Frontier Forest under medium or high threat   ■ Frontier Forest under low or no threat

■ Frontier Forest unassessed for threat   ■ Non-frontier Forest

Preliminary maps of frontier forest boundaries were sent to regional experts. Their corrections were used to create the first worldwide maps of the remaining frontier forests and to assign threat levels to their status as frontier.

Improving information access

The Web site is accessed by hundreds of people daily and is a source of maps, forest information, and links to other resources.

In addition to being a valuable tool for scientists and professional researchers, the site has also become a way for WRI to publicize the severity of deforestation to the media and the public at large.

The Web site also has information about the types of remaining frontier forest, about half of which is boreal forest found in Russia, Canada, and Alaska. Long winters and poor soils in these regions make the land undesirable for farming and have worked to protect these forests.

Very little frontier remains in the temperate regions. Forests once abundant in Europe, China, the United States, Canada, Australia, New Zealand, Chile, and Argentina were cleared for agriculture.

Tropical forests, which until this century remained largely intact, are now being destroyed at the rate of millions of acres yearly for agriculture and wood products.

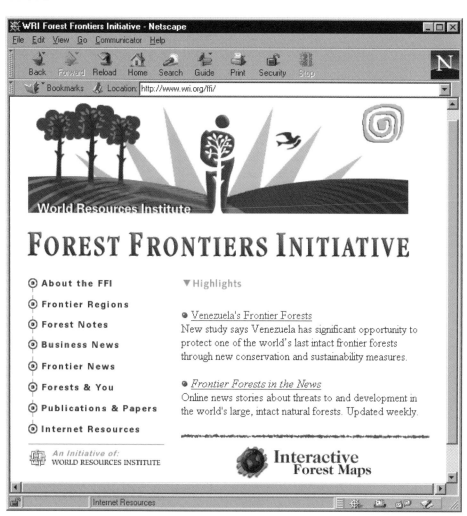

The Forest Frontiers Initiative (FFI) is a five-year, multidisciplinary effort to promote stewardship of the world's last major frontier forests by influencing investment, policy, and public opinion.

Zeroing in on hot spots

The frontier forest maps help the institute's researchers plan how to preserve the world's remaining frontier forests.

They use GIS data and satellite imagery to find "hot spots" of deforestation like northern Burma, northern Vietnam, Laos, and central Africa, where the problem is accelerating. The satellite imagery shows changes in forest cover over time—from year to year, or over the past five or ten years. GIS data depicting newly built roads and highways points to areas that are newly threatened by better accessibility.

The researchers also use GIS to understand regional changes contributing to deforestation. In central Africa, they map the locations of current and planned logging. ArcView GIS software is then used to overlay frontier forest data to show how close those areas are to frontier forests. These maps are used to assign threat levels to vulnerable areas.

Except for the Congo Basin, Africa's frontier forests have largely been destroyed, and most of those remaining are at risk.

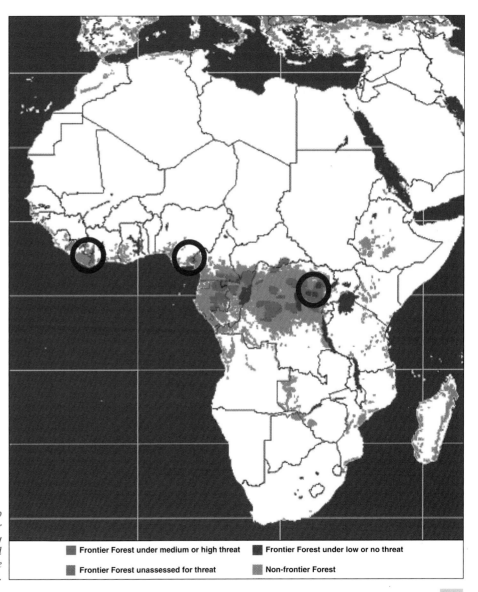

Frontier Forest under medium or high threat

Frontier Forest under low or no threat

Frontier Forest unassessed for threat

Non-frontier Forest

Creating a global care network

The institute is currently starting Global Forest Watch, a follow-up project to the Frontier Forests Initiative that helps regional organizations monitor and report on frontier forest tracts in their areas.

The frontier forest maps are used by researchers to decide where monitoring groups are most needed. Information will be used by these groups to keep track of local changes to frontier forests.

The institute has learned that by using GIS to collect, analyze, and present its data, it can make a compelling case to governments around the world. And by placing maps on its Web site, WRI hopes to make the media and public more aware of threats to the world's resources, including frontier forests.

Regional networks of policymakers, activists, investors, and researchers keep the institute informed of activities in their regions that threaten frontier areas.

Hardware

PCs running Microsoft® Windows NT

Software

ARC/INFO

ArcView GIS

Data

Forests data set (World Conservation Monitoring Centre)

Acknowledgments

Thanks to Daniel Nielsen, GIS analyst at the World Resources Institute.

WORLD RESOURCES INSTITUTE

Chapter 5

Air pollution

Smog can be carried hundreds of miles from its source, causing health and environmental problems on a regional or even global scale. In people, air pollution causes a variety of ailments and can lead to lung damage. It also harms some trees and crops, reducing growth and crop yield, and making them vulnerable to insects and disease.

In this chapter, you'll see how the Environmental Protection Agency (EPA) uses ARC/INFO software to study the effects of air pollution on tree seedlings and to estimate the long-term consequences of smog to forests in the United States. The agency is also making information from many of its programs available over its intranet and the Internet using GIS maps and applications.

Studying the effects of smog

The United States sets emissions levels for automobiles, power plants, and industrial facilities to protect humans and forests from smog, the most visible air pollutant in the country.

An accumulation of gases in the lower portion of the earth's atmosphere, or troposphere, smog is composed mainly of ground-level ozone. Ozone is formed by volatile organic compounds (VOCs) and nitrogen oxides (NOX) in the presence of sunlight and warm temperatures. VOCs (also called hydrocarbons) are found in oil and natural gas, and are released in vehicle emissions and by the evaporation of gasoline and solvents from the petrochemical industry. Like VOCs, NOX are mainly produced by oil and gas, but specifically by the burning of these fuels.

Ozone is the most toxic air pollutant to plants, and its levels near forests concern scientists at the EPA. They're using ARC/INFO software to study the effects of ground-level ozone on forests to predict consequences over the long term.

Analyzing study data

When EPA scientists began studying the effects of air pollution on forests in the eastern United States, their first step was to find out how much ozone reaches the forests. They were particularly interested in how smog affects several native tree species: quaking aspen, black cherry, tulip poplar, sugar maple, red maple, loblolly pine, Virginia pine, eastern white pine, Ponderosa pine, Douglas fir, and alder.

Scientists collected air quality data from the EPA's ozone-monitoring sites, located throughout the United States. Since fewer than 2 percent of these sites are actually in forests, they had to estimate levels for these areas using data collected from nearby monitoring stations.

ARC/INFO software was used to create a grid overlay of the eastern United States, with the study region broken into 20-kilometer-square cells. The scientists could then create a map combining NOX emissions, temperature data, daily cloud cover data, wind direction, elevation, and distance from an emission source. All this data contributes to an understanding of how ozone moves. Depending on a place's topography and the prevailing weather conditions, ground-level ozone migrates away from its source and into other areas.

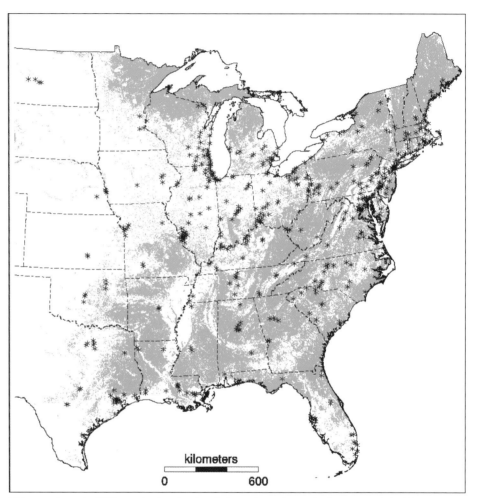

This map shows the EPA's monitoring sites in relation to forested areas in the eastern United States.

Predicting smog movement

The combined data creates a Potential Surface Exposure map, which the EPA uses to assess smog levels at any given location, and to understand where smog forms and in which directions it will probably move.

With this map, the scientists could estimate tree exposure for each cell in the grid during a three-month growing season (July–September) critical to trees. They used ozone levels from two years, 1988 and 1989, for the analysis. 1988 had the highest recorded ozone levels in the decade, while 1989 was a more typical year. The data was used to calculate an average 12-hour (daylight) exposure level.

When the surface exposure map was overlaid with a map of forested areas, it was clear that trees in some cells were exposed to much higher levels of ozone than trees in other cells.

The scientists next wondered how seedlings of different tree species exposed to the same ozone levels would react—were some more sensitive than others? Which ones? And how would those reactions change the forest over time?

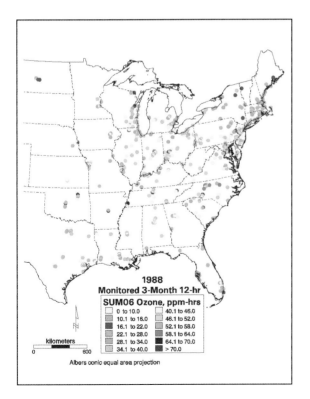
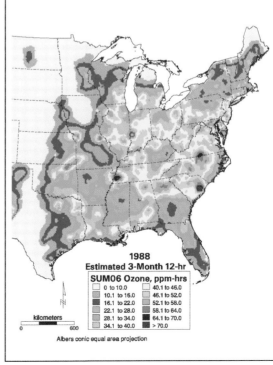

Ozone levels recorded during the growing season were used to create this map of 12-hour average levels for each site location (left). The data was then used to estimate exposure for the entire study area (right).

How smog affects tree growth

To understand how ozone levels affected the growth of individual tree species, the scientists collected seedling growth data from research sites across the United States. The measurements took into account the total biomass (the sum dry weight of roots, stems, branches, and needles) of the growing young trees.

With the GIS, EPA scientists were able to overlay the ozone exposure data for 1988 and 1989, the growth data for the tree seedlings, and the distribution of species to produce a map of biomass loss for individual tree types. Although trees in a given area were subjected to similar smog levels, some were more affected than others.

Aspen and black cherry were the most affected. Loblolly pine, Virginia pine, Douglas fir, and sugar maple were affected the least.

ARC/INFO software was used to mask cities, agricultural areas, and grazing lands from the study area. Satellite imagery was then employed to determine the types of trees in the remaining forested areas. A new grid was created, in which the ranges of individual tree species were divided into 10-kilometer-square cells for further study of the effects of smog on the forests over time.

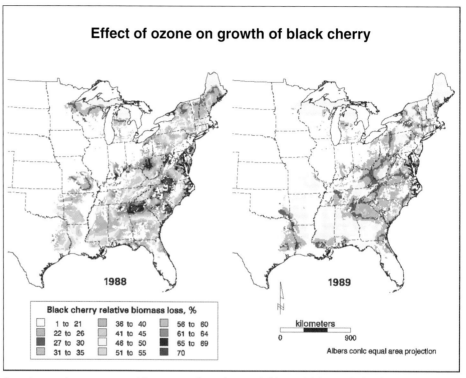

Some tree species are more sensitive to smog than others, and this sensitivity can be heightened by environmental conditions such as drought and changing levels of soil nutrients. Black cherry is a smog-sensitive species. When exposed to 1988 levels of ozone, it lost 30 percent of its biomass over 90 percent of its range. The following year, less than 10 percent biomass loss occurred.

Mapping changes to the forest

The overlay of ozone exposure, growth effect, and species distribution data produced a map showing the extent and magnitude of the smog risk to a particular species in a given year. The scientists were able to predict annual biomass losses from 0 to 33 percent, depending on the sensitivity of the species and the concentration of ozone, which varies from year to year.

According to the forecast, the two most sensitive species, black cherry and aspen, had a biomass loss over 50 percent of their range. Four of the eight species studied had 5 to 12 percent annual biomass loss.

Computer simulations of tree growth that take into account growing conditions, and the influence of these conditions on how each tree species responds to ozone, are now being used to improve the forecasted changes to individual species and the forests as a whole.

The scientists hope this analysis will be useful in predicting the long-term effects of smog on trees from year to year, and in assessing the potential impact of smog on crops, animals, and possibly humans.

Results from the EPA's smog study are used with various predicted ozone exposure scenarios to create maps of how the forests might look in 50 or 100 years—how their boundaries would change and what types of trees would survive.

Environmental data online

Air quality and other EPA data is available at the agency's Web site (www.epa.gov). The award-winning Envirofacts Warehouse contains data and applications for creating and publishing analytical maps of pollution emissions and other environmental hazards.

The EPA's staff uses GIS over the agency's intranet to integrate information in new ways and to adjust regional permits and compliance activities. The Ecological Sensitivity Targeting and Assessment Tool, for example, helps them create maps of contaminants from regulated facilities. They define the facilities and environmental layers (like streams and parks) to view. The application then maps a contaminant's dispersion through both air and water. These maps can reveal potential problems—such as the danger that the contaminant will affect a national wildlife refuge or park—and to alert the agency to the need to protect these resources.

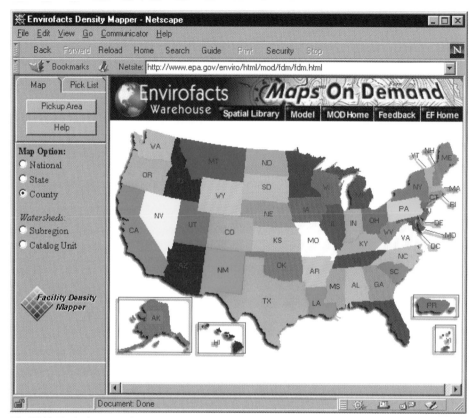

The EPA's Envirofacts Warehouse lets people find air quality data for their region—or the entire United States—over the Internet and create thematic maps.

Hardware

Sun SPARCstation™ workstations

Software

ARC/INFO

TREGRO (tree-growth simulation)

ZELIG 1.0 (forest-stand growth simulator, University of Virginia Environmental Sciences Department)

Data

Forest data (Society of American Foresters and U.S.D.A. Forest Service)

Meteorological data (EarthInfo, Inc., Boulder, Colorado)

Emissions data (National Research Council)

AVHRR satellite imagery (NASA)

Point-source pollution data (ESRI)

AIRS air-quality data (EPA)

Acknowledgments

Thanks to Dr. William E. Hogsett of the United States EPA Environmental Research Laboratory in Corvallis, Oregon.

United States Environmental Protection Agency

Chapter 6

Mining borate ore

Minerals can be found anywhere—from the remote wilderness to crowded cities, from frozen mountains to torrid desert. Valuable deposits may lie in places where digging is inconvenient or a threat to sensitive ecosystems. To be profitable without being disruptive, mining companies and their contractors use GIS to map mineral locations and other pertinent information about terrain, accessibility, and human populations. When they close a mine, they use the same GIS maps to restore the site as nearly as possible to its original state.

In this chapter, you'll see how ArcView GIS and ARC/INFO software help guide a mining operation in California's Mojave Desert, home to several rare or threatened plant and animal species.

Mining borax in the desert

U.S. Borax Inc. (Valencia, California), owned by the world's largest mining concern, Rio Tinto, operates a 1.5-mile-long pit mine that contains borate ore. Located deep in the Mojave Desert near Boron, California, the site includes refining and packaging facilities next to the mine.

Borates are white, crystalline minerals formed in the beds of ancient lakes. According to legend, borates were used by the Egyptians in mummification and by the Romans in glassmaking, but the first historically verifiable use of the minerals was by Arabian gold and silversmiths in the eighth century A.D.

Borates are essential ingredients in many industrial processes, including the manufacture of glass, ceramics, fiberglass insulation, detergents, and flame retardants. Thousands of household products, from barbecue charcoal to brake fluid, also contain borates.

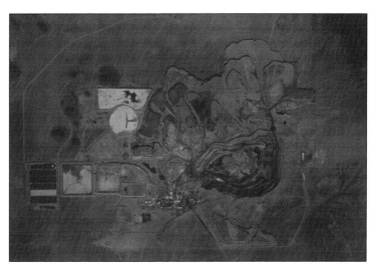

The Borax mine, in the northwestern Mojave Desert, is the largest of its kind on the planet. The mine supplies more than half the world's borates.

Meeting environmental regulations

When the mine first opened in 1927, few people lived nearby, and little was known about the plants and animals that inhabit the Mojave Desert.

While the human population of the region remains small, the mine operators are required to avoid disrupting threatened species such as the desert tortoise. They must know where these species are located in relation to the mine's facilities and roads, and not build new structures or roads where species densities are high.

Mining companies are also required to collect data about how they extract and process minerals, how they handle waste, and what they do to restore the land after closing a mine. This information is reported to regulatory agencies and is used internally to make sure the companies are in compliance with federal and state environmental laws.

Photo courtesy of U.S. Borax Inc.

Heavy equipment is used to mine and transport ore from the pit to the processing facilities. Fully loaded, each truck carries nearly 200 tons of material.

Collecting site data

Borax hired Condor Earth Technologies, Inc., of Sonora, California, to create maps for its planning activities and compliance reports on threatened species, watersheds, and land restoration.

The consultant's first step was to collect data for the 3,000-acre site. From satellite photos and other commercially available GIS data, a base map began to take shape. For the more detailed data that was needed, Condor technicians surveyed the site on foot. Using pen-based portable computers, a proprietary software package called PenMap®, and GPS receivers, they recorded the positions and descriptions of streams, roads, plants, and animals found on the site.

The PenMap system allows operators to enter data (like the location of a stream or tortoise burrow) as they walk around the site and to display the information immediately on a map.

Spatial data collected in this way, along with attribute information, is then converted to shapefile format by the PenMap software. The shapefiles are uploaded to computers running ARC/INFO software for editing and ArcView GIS software for display and analysis.

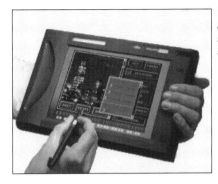

With PenMap, the user collects data and can view it on the screen of the pen computer while still in the field.

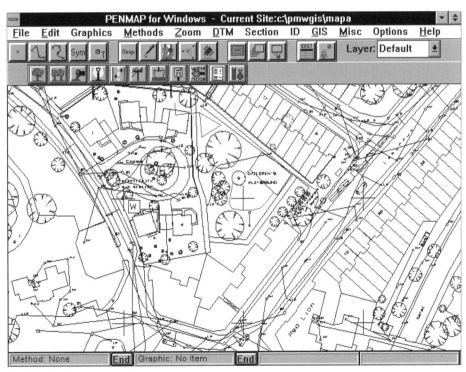

Locating borates underground

The deposit mined by Borax was formed over tens of millions of years through a unique series of geologic events. Ancient lava flows formed a basin that filled with water from underground hot springs. The mineral-rich springs deposited layers of borates on the lake bottom. Over time, the lake dried up, leaving the borates encapsulated in clay and buried under more than 2,000 feet of sand. Finally, seismic activity and erosion brought the deposit closer to the surface, where it could be detected and mined. The site now covers about 1.5 square miles and will ultimately descend as deep as 1,200 feet.

To locate the most promising mining sites in the deposit, the company conducts seismic surveys that involve sending shock waves into the earth to differentiate the layers of rock below the surface. This information is used to develop geologic profiles of the underground layers (so company geologists know how the borate minerals lie). Next, a shaft is drilled and samples are taken to make sure the minerals are present and to determine their quality.

This information is georeferenced and overlaid with the site's topography, botany, and biology to help the company plan its operations.

"Overburden" (the soil, usually sandstone, that covers the borate ore) must be removed to mine the ore body. The question of where to put it is a classic GIS problem. Using ArcView GIS software, analysts overlay active mining sites, known expansion locations, and position of faults and groundwater to determine where the overburden can be relocated without covering areas where future digging might still take place.

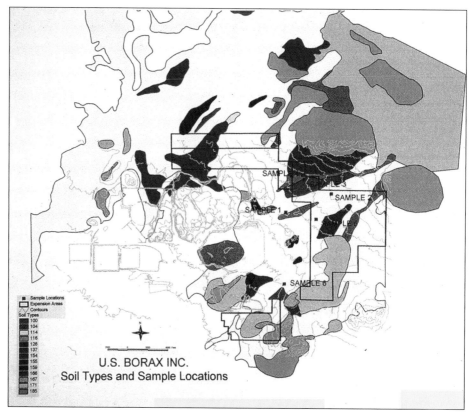

U.S. BORAX INC.
Soil Types and Sample Locations

Soil sampling and seismic surveys are used to build profiles of rocks and water underground so the company can extract minerals efficiently.

Regulatory compliance

The Borax mine is a zero-water discharge facility. In other words, all the water used to process ore is contained throughout the refinery until it's discharged into lined collecting ponds. The material in these ponds evaporates quickly in the desert heat, and the leftover borates are redirected to the refining process.

In the mine, the company uses information from drilling samples to create maps of groundwater and its proximity to underground faults or cracks. These faults break up mineral deposits and serve as groundwater conduits. The maps let the company extract minerals without hitting groundwater or opening a passage that would let groundwater into the mine.

Rain coming onto the site is controlled by a series of catch basins. ArcView GIS software is used to overlay site topography to see where water will drain. Containment facilities are then built at strategic locations. As the company expands the mine operations from one area to the next, the hydrology and topography data is used to plan new runoff basins.

Finally, the company reports its environmental compliance to the government. The Federal Emergency Management Agency (FEMA), in particular, asked for data about the mine's proximity to a nearby floodplain. With ArcView GIS software, the company was able to produce

maps of where it would be working in the deposit in the coming years. It overlaid property boundary information, topography, and drainage to show the agency that its expansion plans (indicated with shading) would be kept well away from the floodplain (on the left side of the map) and included containment facilities for runoff.

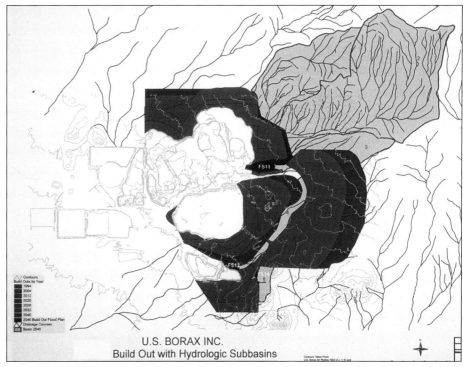

This "build out" map was used to demonstrate that the company would contain water at the site as it expanded and keep work well away from a nearby floodplain.

Land reclamation efforts

The aim of mine reclamation is to re-create the wildlife habitat and promote the return of native animals and birds. To develop a sound reclamation program, the company built more than 32 test plots on land it plans to reclaim—mostly over-burden piles—to find the best way to prepare soil and plant seeds.

The test plots help determine resloping patterns that encourage plant growth: all slopes are bulldozed from the steep 36 degrees at which deposited soil comes to rest to a more stable 18 degrees. The company also tills furrows into the sides of the slopes to capture the limited rain (about 4 inches annually) that falls in the desert.

GIS is used to map and monitor these test sites, and to help mine operators communicate the results of the reclamation efforts within the company.

As Borax continues to operate in the Mojave Desert, GIS data and maps will be used to help the company comply with state and federal regulations and expand its operations in an efficient and environmentally sound manner.

The company's commitment to environmentally sensitive land restoration is one of many corporate programs featured on its Web site (www.borax.com). Borax was honored recently by the State of California for its achievements in solid waste reduction, recycling, and reuse.

Hardware

Pentium-class PCs

Pen computers

Software

ArcView GIS

ARC/INFO

PenMap

Data

Geology

Soils

Hydrology

Topography

Property ownership

Facilities data

Acknowledgments

Thanks to Barry Hillman and Mark Jonas of Condor Earth Technologies, Inc.; Gerry Pepper, Susan Keefe, and Dave Weiss of U.S. Borax Inc.; and Mike Price of ESRI.

**Condor
Earth
Technologies**

Chapter 7

●●●● C l e a n w a t e r

Water pollution does not respect the boundaries of local, state, or federal governments. The tendency of water (and any pollution it may carry) to flow from one pollution-fighting agency's jurisdiction to another has made the problem one of the trickiest of all environmental challenges. But with GIS, many agencies are studying water pollution in a context larger than that of a particular stream segment or side of a lake. Moreover, they are using GIS to share data and, as a result, are working together to help solve their problems.

In this chapter, you'll see how the state of New Jersey uses ARC/INFO and ArcView GIS software to coordinate with other agencies, organizations, and businesses to clean its rivers.

Viewing the watershed

The New Jersey Department of Environmental Protection is charged with protecting and enhancing the state's natural environment, as well as with addressing issues of public health and economic vitality that relate to New Jersey's natural resources.

The department has used ARC/INFO software since the 1980s to assess the water quality of the state's major river basins and to communicate its findings to local governments, community groups, and businesses.

The department studies entire watersheds (areas that drain into a common body of water), so scientists, planners, and regulators can understand all sources of contamination and can work together to restore water quality.

The New Jersey Department of Environmental Protection uses GIS to study the complex dynamics of water pollution.

Collecting field samples

In 1996, scientists from the department's Office of Water Monitoring Management (Bureau of Freshwater and Biological Monitoring) collected samples from the Whippany River, located near Morristown, New Jersey, an urban area. The Whippany watershed receives pollution from a number of sources, including sewage treatment plants, factories, farms, and storm water runoff.

Scientists took samples of the river's sediment and water and measured the concentrations of nutrients, organics, and metals. They also took readings of temperature, dissolved oxygen, and pH. The data was compared to standards set by state and federal agencies to keep the waters clean enough for fish to live in and humans to swim in.

The scientists also collected biological samples, looking in particular for creatures like mayflies, stone flies, and caddis flies, which are sensitive to pollution. If the samples are composed mostly of animals like midges or worms that can tolerate pollution, or if the numbers and types of insects are different than past samples, the scientists are alerted to changes in water quality, changes that could be missed by chemical sampling alone.

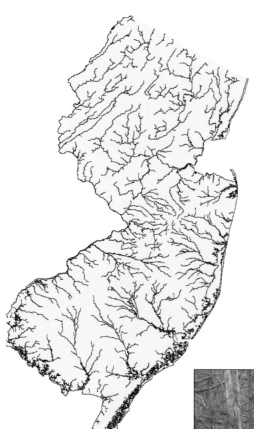

The Office of Water Monitoring Management analyzes the water quality of New Jersey's 1,200 lakes and 6,000 miles of streams and rivers.

The office collects samples from streams and rivers statewide each year to analyze water quality. Data from these samples is entered into ARC/INFO and ArcView GIS software.

Adding the field data to the GIS

As they collected samples, the researchers recorded their exact field positions with global positioning system (GPS) receivers. These positions, along with their associated chemical and biological data, were loaded into ARC/INFO and ArcView GIS software.

To interpret the results of the field data, an analyst used the ArcView GIS system to create a map of the watershed's streams and lakes. Other map layers were added, including topography, roads, watershed boundaries, soils, freshwater wetlands, and open space.

The analyst then added a layer of potential sources of pollution, representing locations, such as factory sites, where pollutants are known to be discharged. The inclusion of land use data made it possible to take into account more diffuse types of pollution, like that from farms and storm runoff.

The analyst queried the GIS to determine the proximity of known sites of contamination, like landfills, to the rivers and streams being monitored.

Finally, the biological sampling data was added and the ArcView GIS system was used to show impairment ratings for stream segments. A stream's impairment rating is a measure of the health hazard it poses to fish and human swimmers.

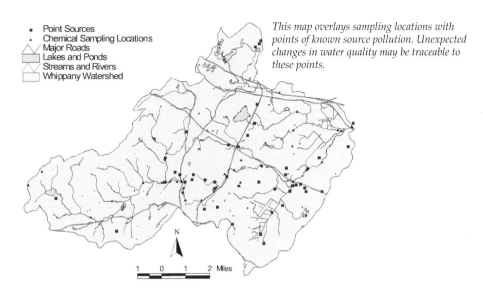

- Point Sources
- Chemical Sampling Locations
- Major Roads
- Lakes and Ponds
- Streams and Rivers
- Whippany Watershed

This map overlays sampling locations with points of known source pollution. Unexpected changes in water quality may be traceable to these points.

1 0 1 2 Miles

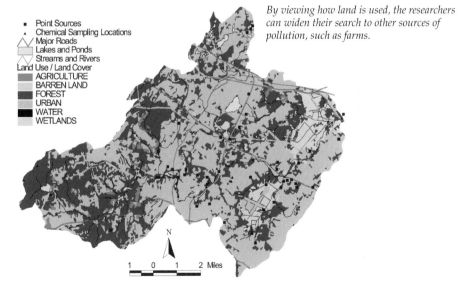

- Point Sources
- Chemical Sampling Locations
- Major Roads
- Lakes and Ponds
- Streams and Rivers

Land Use / Land Cover
- AGRICULTURE
- BARREN LAND
- FOREST
- URBAN
- WATER
- WETLANDS

By viewing how land is used, the researchers can widen their search to other sources of pollution, such as farms.

1 0 1 2 Miles

Modeling contaminants to forecast future levels

The analyst was concerned about the concentration of severely polluted segments (shown in red) along one river, and forwarded the results to the department's planning group.

The planning group can use the data to create models of water quality. By changing the data values being put into the model of specific variables, like the amount of phosphorus coming out of a pipe, they can predict changes to the stream's chemistry and its ability to support insects and other creatures.

The models developed by the agency assign values to river locations based on recorded levels of toxins as well as dissolved oxygen, phosphorous, and solid wastes. The planning group can use this data to forecast how small changes in chemistry may, if allowed to continue, affect the river in the coming years.

The model they created with data about these severely impaired streams indicates that this toxin, copper, if allowed to continue, would spread quickly and impair more segments of this river, killing fish and wildlife. Now the department's environmental regulation group can take action, visiting farms, landfills, and factories along this segment to check for leaking storage tanks or runoff. If these inspections are inconclusive, the office can conduct additional water sampling to

help its regulators find out where contaminants are entering the stream. With this information, they can usually identify a polluter. This time, a storage tank at a nearby factory was found to be the source.

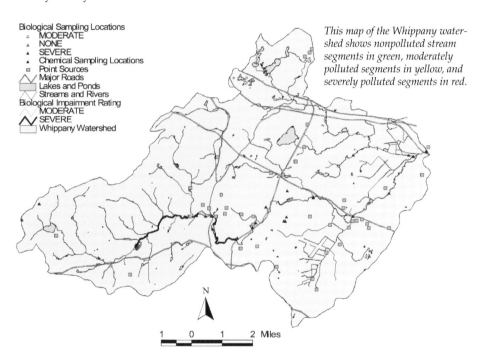

Biological Sampling Locations
△ MODERATE
▲ NONE
▲ SEVERE
▲ Chemical Sampling Locations
▫ Point Sources
⋀ Major Roads
▢ Lakes and Ponds
Streams and Rivers
Biological Impairment Rating
MODERATE
⋀ SEVERE
Whippany Watershed

This map of the Whippany watershed shows nonpolluted stream segments in green, moderately polluted segments in yellow, and severely polluted segments in red.

N

1   0   1   2 Miles

Using GIS to educate

The department's planning and environmental regulation groups use the results of the analysis to cooperate with local agencies to locate previously unknown polluters and improve the area.

They continue to test water during this process to ensure that the stream's condition is improving and contaminants have been correctly identified.

The GIS maps and data are also used when the office meets with local health officials, planning groups, businesses that discharge regulated waste, and citizens' groups.

Together, these organizations identify water quality issues and potential areas of investigation and set limits on chemicals entering the stream system.

The department continues sampling rivers in areas of impairment to monitor progress.

Publishing data for the public

The GIS data used in the watershed program has been combined with other data layers and published on a CD–ROM available to the public for $30. The CD includes a limited-use version of ArcView GIS software, statewide data in an ArcView GIS-compatible format, narrated demos, and a directory of GIS references.

With this CD, home owners can research the locations of toxic sites near their homes or produce maps for community meetings.

In 1998, portions of this data will be available on the Internet. An ArcView GIS application will let users enter an address (or point to a map location) to view a property and nearby toxic sites, water sources, political boundaries, and more.

By publishing its watershed data on the Internet and encouraging communities to look at distant contributors to local water pollution problems, New Jersey has fostered a climate of change that's already measurable in cleaner water. Thirty-five percent of the state's 3,815 stream miles fully support aquatic life, while another 53 percent partially support it. New Jersey's goal is to have 50 percent of its streams supporting healthy aquatic life by 2005.

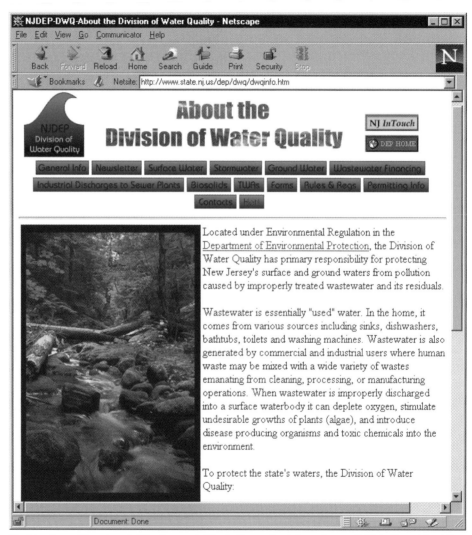

The New Jersey Department of Environmental Protection makes information available to the public at its Web site (www.state.nj.us/dep).

Hardware

Pentium-based PCs

Sun SPARCstation workstations running Solaris™

Software

ARC/INFO

ARCPLOT

ArcView GIS

Data

Environmental (geology, soil types, and land use)

Areas of concern (freshwater wetlands, open spaces, water supply intakes, community drinking water wells)

Potential sources of pollution (surface and groundwater wastewater discharge, hazardous waste sites, and facilities on the EPA's Toxic Release Inventory)

Information about chemical and biological water quality

Acknowledgments

Thanks to Paul Morton and Patricia Cummins of the New Jersey Department of Environmental Protection.

Chapter 8

Reclaiming brownfields

Many urban areas have vacant properties that are suspected or known to be contaminated from former industrial use or illegal dumping. These properties, known as "brownfields," blight communities and undermine the property values of surrounding areas. Developers and investors are understandably wary of buying places like this, places that they know will be costly to improve. Some local governments are using GIS to locate brownfields and to help return them to productive use.

In this chapter, you'll see how the Mayor's Office of Environmental Coordination in New York City uses ArcView GIS software to measure the problem and communicate its findings to the development, investment, and environmental communities as well as to the affected neighborhoods.

New York takes stock

In the late 1890s, New York City was one of the world's leading industrial centers. Fed by European immigration, its rapidly expanding population worked in hundreds of factories in Manhattan, Brooklyn, Queens, and the Bronx. Located mainly along the waterfront, these factories produced clothing, foods, machinery, ships, instruments, furniture, and coal and petroleum products. They also produced waste—waste that was sometimes dumped at the site or buried underground.

Today, many of these former factory sites are abandoned properties. Collectively, they are known as brownfields—not the unkindest name imaginable for these potentially hazardous eyesores.

The Mayor's Office of Environmental Coordination heads up the city's Brownfields Initiative, which directs the city to work with relevant public agencies, property owners, developers, and local community boards to gather information on brownfields and analyze it. The data is an essential part of the effort to redevelop the land and revitalize local economies.

Decades after the decline of major industry in New York, local officials are still dealing with the aftereffects.

Building a brownfields system

The federal Land Recycling Act of 1997, commonly referred to as the Brownfields Act, ordered the Environmental Protection Agency (EPA) to help local governments redevelop contaminated sites by reducing liability to prospective purchasers and reducing, if not removing, significant risks to human health and the environment.

New York City is one of 157 states, cities, towns, or tribes receiving a $200,000 grant from the EPA. The grant requires the city and its project partners in the Brownfields Initiative to select for cleanup five pilot sites that illustrate the constraints and opportunities associated with New York City's brownfields.

With thousands of sites to choose from, the Mayor's Office of Environmental Coordination needed a way to evaluate brownfields properties and neighborhoods. The criteria were to select one site from each borough, and have at least one of the five sites located in a federally defined Empowerment Zone in northern Manhattan or the South Bronx. They also wanted each site to have different types of contamination and be slated for different types of uses—residential, commercial, open space. They also had to be available—either owned by the city or by a private owner willing to participate in the program.

The city contracted HydroQual, Inc., of Mahwah, New Jersey, to map the project digitally. With ArcView GIS software, the consultants created a base map by linking tax records with parcel files from the planning department. To this was added census, streets, and zoning data. The city began using the system in 1997.

GIS helps the city determine which properties can be redeveloped. At some sites, cleanup will be completed before the property is transferred to new owners. At others, cleanup may take place simultaneously with construction and redevelopment activities.

Characterizing potential sites

Analysts from the city queried the GIS for vacant industrially zoned parcels. They found more than 6,600 properties fitting this description, totaling 3,300 acres. While a number of these sites were already clean and slated for redevelopment, the search provided a starting point for the project.

They then found which brownfields were in federal Empowerment Zones or state Economic Development Zones, which might qualify them for additional funding for cleanup.

In total, the city's brownfields had an appraised value of hundreds of millions of dollars. The potential revenue from the sale of city-owned properties and from newly collectible taxes on redeveloped sites would be significant.

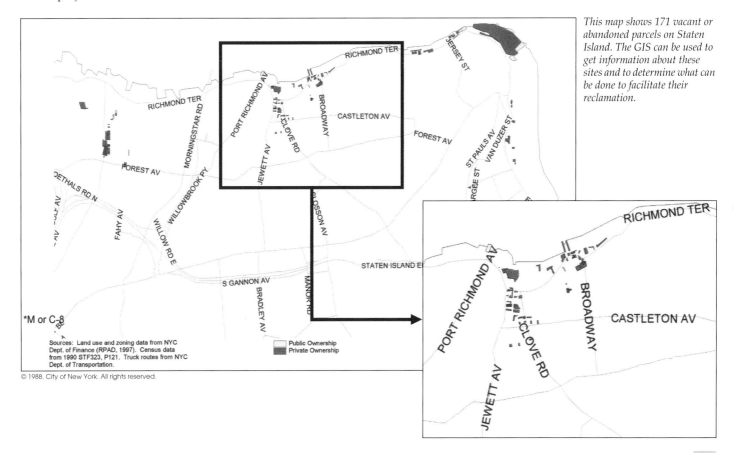

This map shows 171 vacant or abandoned parcels on Staten Island. The GIS can be used to get information about these sites and to determine what can be done to facilitate their reclamation.

Sources: Land use and zoning data from NYC Dept. of Finance (RPAD, 1997). Census data from 1990 STF323, P121. Truck routes from NYC Dept. of Transportation.

Public Ownership
Private Ownership

*M or C-8

Increasing community awareness

Using ArcView GIS software to overlay local political boundaries, the analysts found that most brownfields were situated in 26 of the city's 59 community board districts. Knowing the location of the brownfields helped the mayor's office determine which local groups to approach about the project.

By meeting with local groups, the mayor's office was better able to enlist community help in prioritizing the different sites. The local groups were also able to help identify development, investment, and environmental issues associated with these sites.

Before community meetings, the analysts used the GIS to find information about the brownfields in each region. The system provides data on property ownership, land use and zoning patterns, property size, distance to major arterial rail lines and waterways, and demographics. Having this data on hand allowed city officials to be responsive to questions raised by those living in the affected area.

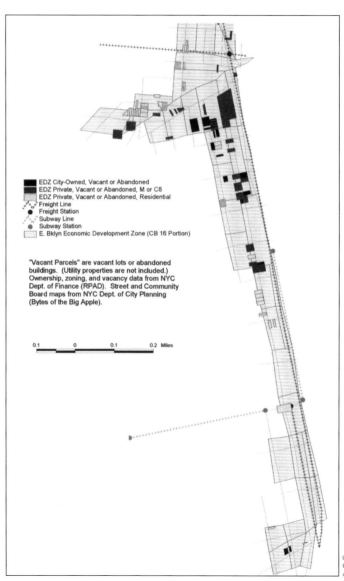

EDZ City-Owned, Vacant or Abandoned
EDZ Private, Vacant or Abandoned, M or C8
EDZ Private, Vacant or Abandoned, Residential
Freight Line
Freight Station
Subway Line
Subway Station
E. Bklyn Economic Development Zone (CB 16 Portion)

"Vacant Parcels" are vacant lots or abandoned buildings. (Utility properties are not included.) Ownership, zoning, and vacancy data from NYC Dept. of Finance (RPAD). Street and Community Board maps from NYC Dept. of City Planning (Bytes of the Big Apple).

0.1   0   0.1   0.2 Miles

During community meetings, the analysts use ArcView GIS software running on a laptop to identify brownfields, discuss which sites are most important to the community for redevelopment, and update databases.

Analyzing site data

The GIS was also used to evaluate brown-fields for EPA pilot sites and/or sites suitable for State Bond Act funding. One important factor to consider was the type of contamination at a site. Historically, certain industries were concentrated in particular parts of the city. Some of these industries produced more hazardous waste than others and left certain contaminants behind. Some of these contaminants can now be cleaned from soils with new technologies. Properties with these specific contaminants were considered desirable as pilots.

Maps were also generated of parcel ownership and proximity of properties to federal Empowerment Zones and Economic Development Zones—factors important for selecting the five final sites.

Another, more direct, benefit of GIS is its use as a marketing tool. Once a property has been cleaned, or found to be uncontaminated, the system can analyze its proximity to transit lines or building types—features that may make it more attractive to potential buyers.

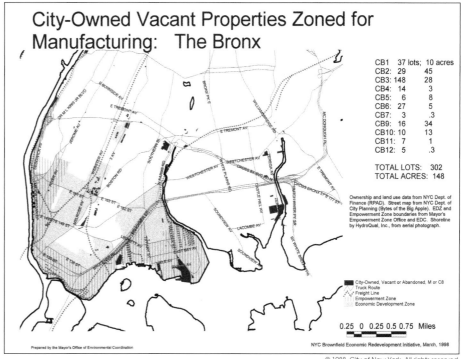

City-Owned Vacant Properties Zoned for Manufacturing: The Bronx

CB1	37 lots;	10 acres
CB2:	29	45
CB3:	148	28
CB4:	14	3
CB5:	6	8
CB6:	27	5
CB7:	3	.3
CB9:	16	34
CB10:	10	13
CB11:	7	1
CB12:	5	.3

TOTAL LOTS: 302
TOTAL ACRES: 148

Ownership and land use data from NYC Dept. of Finance (RPAD). Street map from NYC Dept. of City Planning (Bytes of the Big Apple). EDZ and Empowerment Zone boundaries from Mayor's Empowerment Zone Office and EDC. Shoreline by HydroQual, Inc., from aerial photograph.

■ City-Owned, Vacant or Abandoned, M or C8
Truck Route
Freight Line
Empowerment Zone
Economic Development Zone

0.25 0 0.25 0.5 0.75 Miles

Prepared by the Mayor's Office of Environmental Coordination

NYC Brownfield Economic Redevelopment Initiative, March, 1998

By overlaying the location of known brownfields, arterials, and current Empowerment Zones (neighborhoods targeted for new business and investment), the analysts can see which brownfields in a borough are located within an Empowerment Zone, and which are not.

Expanding the GIS database

With the brownfields database nearly complete, the city will select its five pilot sites by the end of 1998.

In the future, the city will use aerial photography as a backdrop to the GIS data to make the positions of parcels, including brownfields, more accurate. This will make the information of greater value for local planning projects and help the mayor's staff better support economic incentive programs, infrastructure improvements, and changes in zoning and land use controls.

The office also hopes to add other resources to the database. Paper maps and historic data can help show how brownfield properties were used during the late 1800s and early 1900s—the period when many of these sites were contaminated.

Parcel by parcel, New York City is reclaiming its brownfields, improving local neighborhoods and the city's economy, and becoming a model for other governments hoping to use GIS in land-recycling programs.

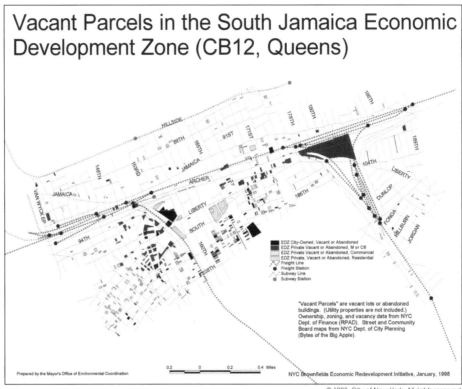

Vacant Parcels in the South Jamaica Economic Development Zone (CB12, Queens)

EDZ City-Owned, Vacant or Abandoned
EDZ Private Vacant or Abandoned, M or C8
EDZ Private Vacant or Abandoned, Commercial
EDZ Private, Vacant or Abandoned, Residential
● Freight Station
⋯ Freight Line
⋯ Subway Line
◎ Subway Station

"Vacant Parcels" are vacant lots or abandoned buildings. (Utility properties are not included.) Ownership, zoning, and vacancy data from NYC Dept. of Finance (RPAD). Street and Community Board maps from NYC Dept. of City Planning (Bytes of the Big Apple).

0.2   0   0.2   0.4 Miles

Prepared by the Mayor's Office of Environmental Coordination   NYC Brownfields Economic Redevelopment Initiative, January, 1998

Parcels in Economic Development Zones may qualify for tax breaks from the state. With this map, the city can determine which properties may qualify and use that information to choose its pilot sites for the EPA grant.

Hardware

Pentium-based PCs

Software

ArcView GIS

Data

The City of New York provided all of the following data:

Parcel maps

Tax lots

Street/Centerline/Transit data

Truck route data

Political boundaries

Demographic data

Shoreline data

Health center districts

Emergency services districts

School districts

Spill records

Hazardous waste sites

EPA Superfund sites

Chemical/Oil storage facilities

Historic and current land use data

Acknowledgments

Thanks to Director Annette M. Barbaccia and Deputy Director Ben Miller of the New York City Mayor's Office of Environmental Coordination, Gary Ostoff of HydroQual, Inc., and David LaShell of ESRI.

City of New York

HydroQual, Inc.
Environmental Engineers and Scientists

Chapter 9

Coastal protection

Coastal regions, with their temperate climates and access to the sea, have historically supported greater levels of human activity than more remote inland areas, with unfortunate consequences for the environment. Loose soil from construction silts up the coast, reshapes the sea bottom, and smothers plants and shellfish. Factory pollution contaminates the water in the streams and rivers that carry it and in the bays and inlets where it comes to rest.

In this chapter, you'll see how the state of Delaware uses ArcView GIS software to guide construction projects and help protect the ocean.

Stopping pollution at its source

In Delaware, where no point of land is farther than 8 miles from tidal waters, development projects can have unforeseen consequences for the coast. Over the past 20 years, many areas have seen rapid commercial and residential building. As a result, nearby rivers and streams have carried pollution and loosened soil into the Delaware and Chesapeake bays and the Atlantic Ocean.

This runoff has taken its toll in the loss of marine life and plants, and in the declining numbers of fish available for commercial and recreational fishing.

In an effort to study the problem from a geographic perspective, scientists from the Delaware Coastal Management Program have created a GIS data warehouse of coastal and marine features for Delaware's 260 miles of coast to share with local governments.

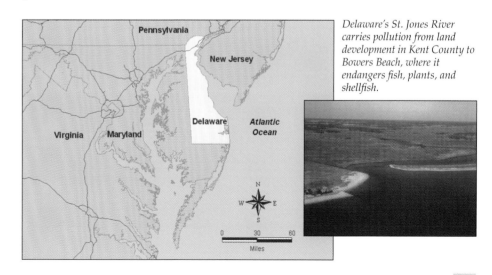

Delaware's St. Jones River carries pollution from land development in Kent County to Bowers Beach, where it endangers fish, plants, and shellfish.

Building an ArcView GIS application

Kent County, the middle one of three in the state, used this data to develop an ArcView GIS application called COMPAS Delaware: Kent County Resource Protection Module. COMPAS is an acronym for Coastal Ocean Mapping Planning and Assessment System. The program runs on PCs and includes data from several resource management agencies.

The county's land use is diverse. Its western part, bordering Maryland, is mainly agricultural. The middle part, where most of the commercial development is located, contains important urban centers and roads. The eastern part, extending to Delaware Bay, contains extensive wetlands, forests, and endangered species, along with some agricultural land.

Development in any area affects the coastal regions, so planners are using the GIS to decide where development should and shouldn't take place.

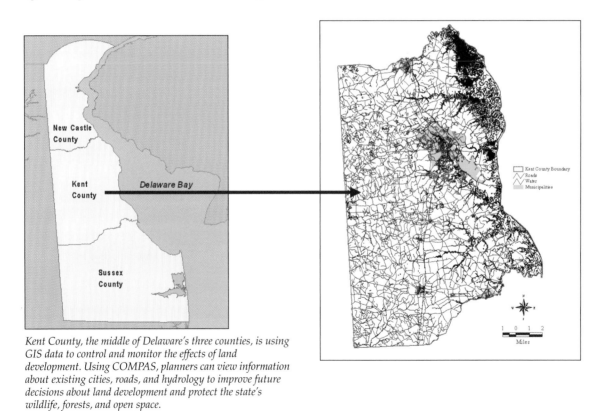

Kent County, the middle of Delaware's three counties, is using GIS data to control and monitor the effects of land development. Using COMPAS, planners can view information about existing cities, roads, and hydrology to improve future decisions about land development and protect the state's wildlife, forests, and open space.

Spotting land use conflicts

Before issuing building permits, Kent County planners research information about the parcel targeted for development and its surrounding area. This process used to take weeks. Using COMPAS, they are able to conduct environmental research in more depth and more quickly than before and provide answers to developers within days.

The parcel identification number or a place name is entered to display the location on a map. The system allows them to overlay as many as 50 layers of data, from agricultural preservation districts and wildlife information to solid waste locations.

The application automatically highlights possible obstacles to developing a selected parcel. On this map, the planner sees the location of the parcel in relation to noise levels from a nearby air base. The area between the purple (75-decibel) and red (80-decibel) lines is unsuitable for some types of development.

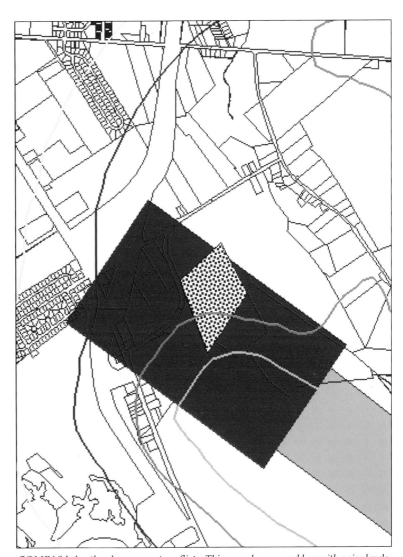

COMPAS helps the planners spot conflicts. This map shows a problem with noise levels from a local air base.

Guiding development to safer areas

The planners will use results from system queries to steer growth into areas that can support it, while preventing contamination of rivers or destruction of resources.

ArcView GIS software is used to create a 2-mile buffer zone around existing infrastructure like roads and sewer pumping stations. The buffered areas, within which development is to be contained, are then merged, and data on planned roads and highways is overlaid to find potential growth zones.

The GIS identifies land that has already been developed, and land that shouldn't be developed, such as wetlands and protected historical sites.

The resulting map shows growth zones, outlined here in red, where existing or planned infrastructure can support high-density housing (on the order of eight homes per acre). Development in these areas will not adversely affect streams, wetlands, endangered species, state historical sites, forests, or other natural resources.

Local governments offer incentives to guide growth into these areas. Conversely, restrictions are placed on development projects outside these zones.

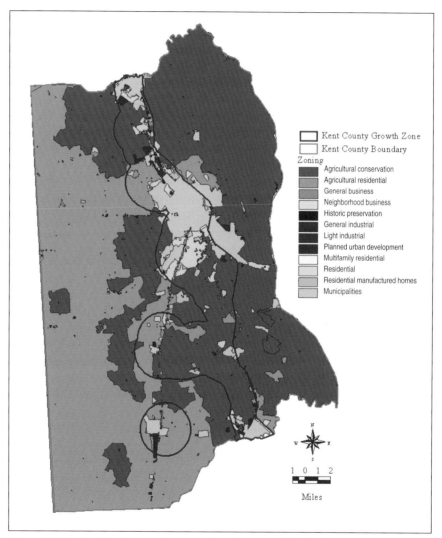

The application is used to plan growth zones, where the county feels development can be supported by planned or existing utility lines, sewers, and roads and won't damage the local environment or the coast.

Protecting resources

Forests provide habitat for endangered species, help clean the air, and prevent soil erosion. Delaware's interior forests, especially those of more than 25 acres in size, are the most critical. Removal of trees from these forests for development or agriculture adds to the state's coastal pollution, habitat loss, and smoggy air.

To better understand this issue, the planners use COMPAS to map the critical 25-acre stand of trees.

Using this information, analysts can begin to understand the impact of developing this land. By overlaying hydrology, they can see where soils loosened by development could enter streams and rivers and travel to the coast. The department uses maps like these to make decisions on development projects that would reduce the state's forested land and add to coastal pollution. While some of this land may ultimately be developed, the GIS keeps a running tally of its loss and helps planners set aside the most important areas, where sensitive streams and habitat are abundant.

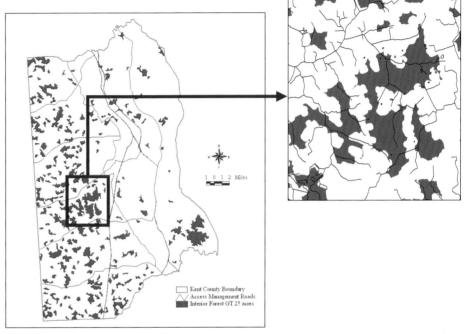

This map helps planners understand how soil runoff from the development of forested land might enter streams and eventually travel to the coast.

Mapping endangered species densities

To protect rare and endangered species in Delaware's remaining forests, the planners need a clear picture of where these plants and animals are found.

Using data from the state's Natural Heritage Inventory, they employed ArcView GIS software's ArcView Spatial Analyst extension. From the locations of various species measured at each of the sample sites (shown as dots), this software extension produced a continuous surface map showing estimated species density per square mile. (Darker colors indicate higher densities.)

The map identifies a region in western Kent County with many rare and endangered species and some historical sites. This data can be used to keep development away from sensitive areas.

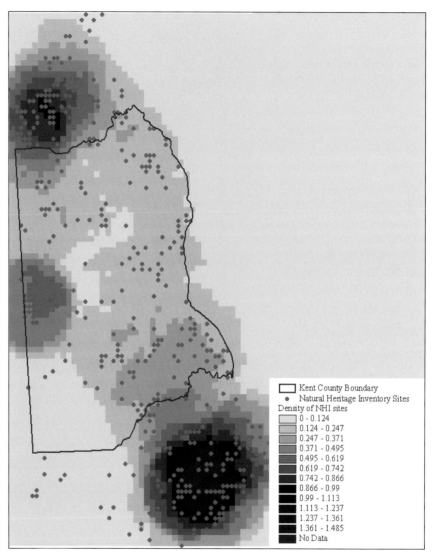

Kent County Boundary
• Natural Heritage Inventory Sites
Density of NHI sites
0 - 0.124
0.124 - 0.247
0.247 - 0.371
0.371 - 0.495
0.495 - 0.619
0.619 - 0.742
0.742 - 0.866
0.866 - 0.99
0.99 - 1.113
1.113 - 1.237
1.237 - 1.361
1.361 - 1.485
No Data

This map showing densities of rare and endangered species in and around Kent County helps planners decide where to allow future residential or commercial development.

Planning at the local level

Planners will also use the GIS to refine the county's Comprehensive Land Use Plan, which determines how much land will be built on, how much will be set aside for recreation or future growth, and how much will be protected. This plan is updated every five years.

On this map (a close-up of the central part of the county), the planner overlays current land use and a proposed development tract (yellow) to see how much open space will be lost if the land is developed.

Kent County has not used the GIS long enough to measure environmental improvements, but planners say it's helping them guide development to reduce future water pollution and contamination of the coast.

Eventually, they hope to make GIS data and some COMPAS mapping applications available over the Internet so other local governments, conservation groups, and the community can understand the link between inland soil erosion and ecology-disrupting sediment buildup in bays and inlets.

Parcel to be Developed
1992 Land Use
Residential
Commercial
Industrial
Transportation
Other Mixed Urban
Recreational
Agricultural
Rangeland
Forestland
Water
Wetlands
Barren

Highlighted Parcel (yellow cross-hatch) is to be developed. It is currently Agricultural and development of it would respresent a significant loss of open, agricultural land.

The yellow highlighted parcel is currently used for agriculture. Developing it would represent a loss of open agricultural land in an area that has already been developed for commercial use and residential housing.

Hardware

Pentium-class PCs

Software

ArcView GIS

ArcView Spatial Analyst

Data

One-hundred-year floodplains

Agricultural preservation districts

Aerial photographs

Airport approach zones

Census tracts

Permit data

County protected areas

Emergency services

Hazardous materials/Toxic release
inventory data

Forests

Greenways

Groundwater recharge

Growth zones

Historic sites

Natural heritage inventory

Parcels

Open space

Railroads/Roads

Census data

Scenic areas

Schools

Sewers

Soils

Topology

State resource protection areas

Wetlands

Acknowledgments

Thanks to David Carter and Miriam Lynam of the Delaware Coastal Management Program, Department of Natural Resources and Environmental Control; and Connie Holland and Kevin Coyle of the Kent County Department of Planning.

Chapter 10

Forests and wildfires

The forest is a complex ecosystem. Making sure all of its rivers, streams, plants, and wildlife remain healthy, and can successfully coexist with human activities, is an immense task. More and more often, foresters are finding that GIS makes this task easier. It helps them organize and relate information from different agencies, and use this information to make better decisions about forest resources.

In this chapter, you'll see how Oregon's foresters use ArcView GIS, ARC/INFO, and MapObjects™ software to protect the state's forests from fire, disease, and people.

Tracking fires and equipment

The Oregon Department of Forestry protects 16 million acres of private, state, and federal forest land. From early July through late September, thousands of wildfires occur in the state, some started by lightning, others by careless campers or children playing with matches. These fires cost an estimated $20–$30 million annually in manpower, resources, and fire-fighting equipment. At the season's zenith, a dozen fires might be burning at once.

It's no wonder that during the off-season, planners at the Oregon Department of Forestry use GIS software to analyze fires from previous seasons and plan for the coming year.

The technology helps them do several things: pinpoint areas where fires have been frequent, identify fire hazards, coordinate on-site fire-fighting resources, and deploy fire-fighting equipment and crews more effectively.

In Oregon, foresters rely on GIS technology to support decision making and deploy fire-fighting resources more effectively.

Understanding how preventable fires start

The Southwest Oregon District in Medford protects 1.8 million acres in Jackson and Josephine counties, an area that has about 250 forest fires a year.

The district maintains historical dBASE® data and makes it available over a local area network for off-season planning. This data includes when and where fires occurred, how much acreage they burned, what they cost, and how they were started. The last category is significant because it helps the district take action to prevent fires caused by humans. These fires typically burn near structures and endanger lives.

Because the tabular dBASE files include latitude/longitude values for the location where each fire started, it's possible to geocode these locations with ArcView GIS software—that is, to position them accurately on a map. The example at the right shows the locations of fires over the past several years and identifies their causes.

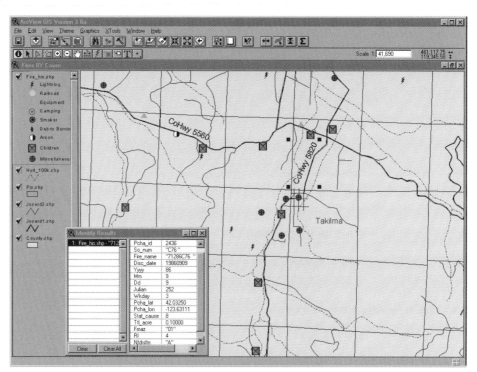

On this stretch of Highway 5820, near the town of Takilma, there have been several fires caused by children. Identifying trends like this helps the district allocate funds for fire education programs in local schools.

Coordinating response to wildfires

In addition to its value as a tool for analyzing historical data, GIS has more immediate fire-fighting applications. It's used along with global positioning system (GPS) receivers to provide accurate positions of fire lines and ground crews, structures, valuable timber, and water sources.

The GPS consists of 24 satellites that send radio signals to mobile receivers on the ground, fixing the receiver's latitude and longitude, as well as its speed and direction. This information can be used in the GIS for visual display and real-time tracking.

During an electrical storm, the positions of lightning strikes are provided to the state by a private data source. The district then sends a spotter plane with a GPS receiver to locate fires that might have started. The location of the wildfire and an estimate of its size and direction are transmitted by the spotter plane to the dispatcher at the district office. The GIS shows the fire's location on a map and identifies the best routes to it.

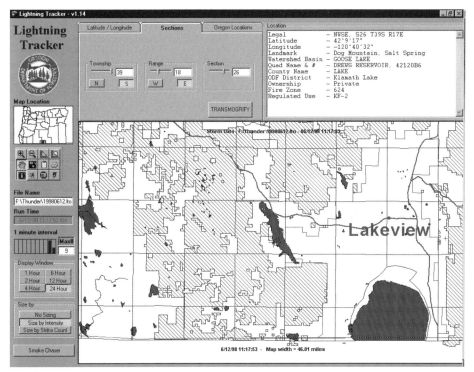

The Lightning Tracker application runs in the department's dispatch centers. During storms, information about lightning strikes is received at the main office in Salem. The data is relayed to the dispatch centers and displayed on a map.

Planning for next season

As soon as crews are dispatched, the district's GIS is used to track their location, indicate other available crews and trucks, and create maps of access roads and ponds or lakes where trucks and helicopters can get water.

The system is also used to direct firefighting activities. Once the crews are at the fire, they are often bound by land management restrictions. Certain environmentally critical areas under the jurisdiction of the Bureau of Land Management can't be bulldozed. In these areas, the crews have to get to the fire on existing roads and stop it without making fire lines.

After each fire, data about its location and the acreage burned is added to the database and evaluated to prepare for the next one.

Medford's foresters use ArcView GIS software to create computer models showing response times to past fires and considering the impact of alternative storage sites for equipment. These models help the department locate resources for improved response times in coming years.

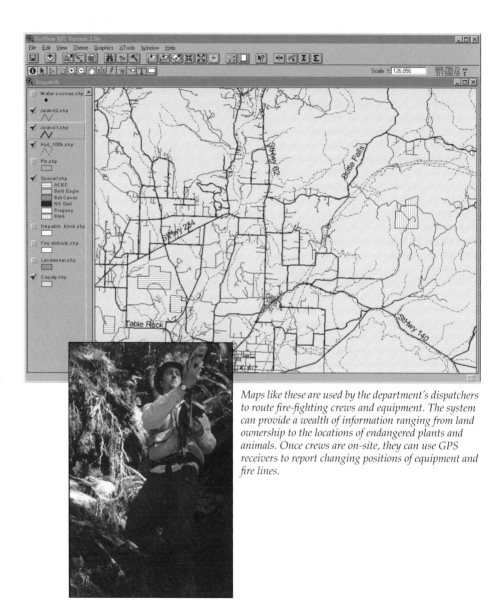

Maps like these are used by the department's dispatchers to route fire-fighting crews and equipment. The system can provide a wealth of information ranging from land ownership to the locations of endangered plants and animals. Once crews are on-site, they can use GPS receivers to report changing positions of equipment and fire lines.

Managing forest habitat

In addition to its fire-fighting activities, the department has other programs that concentrate on forest management—making sure that trees and wildlife remain healthy and safe from human activities.

The Western Lane District in Veneta is the first to use a MapObjects application to track "Notifications of Operations." These are formal requests by members of the public to perform activities like planting, thinning, spraying herbicide, or harvesting trees on private forest land. The district receives about 2,000 such requests a year. Its foresters can review them for compliance with environmental regulations and respond within 15 days.

The application is used, for example, when a private landowner files a notification to harvest trees. A forester reviews the notification by examining pertinent spatial information about streams, fish, wetlands, domestic water supplies, endangered species locations, scenic highways, and steep slopes. Soils data is overlaid to check for potential erosion into streams.

The application automatically queries other map layers, so the forester can view conflicting activities, overlapping requests, and other information about the site.

The request can then be approved, denied, or kept pending until the forester can schedule an on-site inspection. Some analyses reveal no conflicts and the application is approved in a single day.

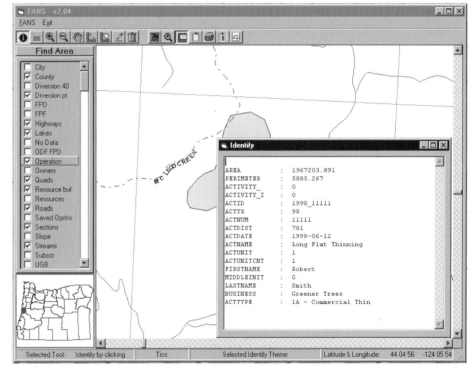

This application, built with MapObjects software, helps the district's foresters review and respond to formal requests to perform activities on forested land. The foresters check for conflicting activities, such as timber harvesting and moss gathering, to ensure that a proper balance is maintained.

Tree disease and other hazards

State forest land managers also use ArcView GIS and GPS systems to track insect and disease infestations and to schedule treatments.

Root rot is a group of plant diseases that travels from root to root, mainly in conifers like the Douglas fir. In the field, foresters use GPS receivers to record the position of infected trees within stands. These positions are then uploaded to the GIS, where buffer zones are drawn around areas within affected stands.

This map shows areas with infected trees. The foresters use it to schedule reforestation, during which they will replace the diseased trees with red alder seedlings. Red alder is a species highly tolerant to root rot.

Data collected about insects that damage timber, like the western spruce budworm and Douglas fir beetle, can also be entered into the GIS and used to plan.

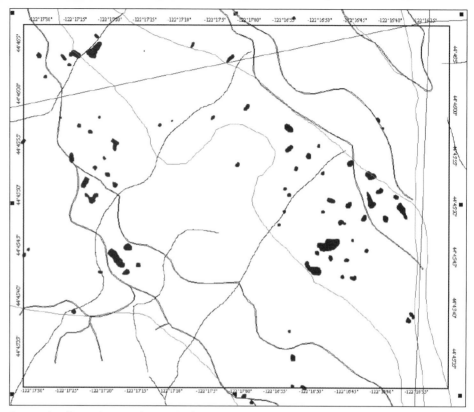

Root rot is a disease that must be contained as soon as possible, since it travels quickly and damages valuable timber. Using GPS equipment, foresters enter the locations of infected trees so they can be replaced with seedlings of a resistant species like red alder.

Monitoring timber sales

Other districts are preparing their data to use with ArcView GIS software and the department's MapObjects applications.

Foresters in many districts are using GPS receivers to remap the locations of roads and to survey section corners of public lands.

Data points collected in the field are loaded into ArcView GIS and ARC/INFO software and used to correct the district's base map.

In this example, the perimeter of a timber sale is shown. Before the data was corrected, the district's maps were difficult to use. Since the agency collects revenue from timber sales based on acreage, the accuracy of the maps is important to the department and to third parties like lumber companies.

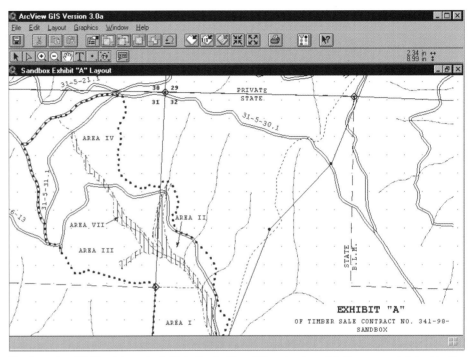

Newly captured field points are loaded into the ArcView GIS system to update the maps maintained by the districts. This map is used to define where timber can be harvested.

Opening data to the public

Oregon's state foresters aren't the only ones tapping into the department's online data. The Oregon Latitude Longitude Locator is a MapObjects application developed by the department that provides GIS data and applications over the Internet (www.odf.state.or.us).

Users enter place names or latitude/longitude positions to view detailed maps of an area. By pointing to on-screen menu choices, they can view data layers such as forest protection districts, USGS quad sheets, watersheds, seed zone/average rainfall, and elevation bands.

Since some trees thrive only in particular seed zones or at certain elevations, this data is particularly useful to owners of private forests for planning seedling purchases and reforestation.

The data could also be used to plan recreational outings or to contact the appropriate district to report a fire or illegal tree cutting.

As additional GIS applications are launched on the department's intranet, Oregon's foresters will have the data and mapping tools necessary to respond faster to wildfires, disease outbreaks, and insect infestations.

The department's forestry management decisions will not only be made faster, but with more and better information, ensuring better care of the entire forest ecosystem.

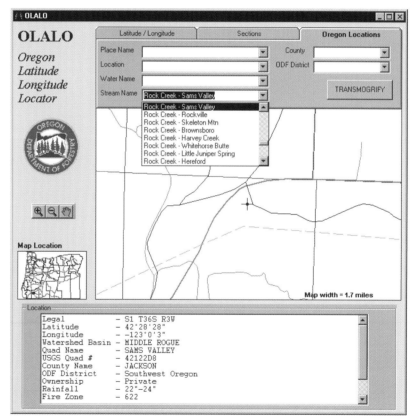

The department is making its interactive maps and information available over the Internet with MapObjects software.

Hardware

Pentium-class PCs supported by UNIX workstations running Sun Solaris

Software

MapObjects

ArcExplorer™

ArcView GIS

ArcCAD®

ARC/INFO

Data

For state forest lands: 1:12,000 streams and roads and 1-meter color orthophotography.

For the rest of the state: the Oregon Department of Forestry is working closely with other agencies to develop 1:24,000-base data and 1-meter black-and-white orthophotography.

Acknowledgments

Thanks to Emmor Nile, Laura Lakey, Vince Pyle, Robert Nall, and Jim Wolf of the Oregon Department of Forestry.

Chapter 11

• • • • • Endangered species

In the Pacific Northwest, overfishing and changes to streams caused by runoff from agriculture and development have endangered many species of salmon and trout. These fish are vital to the ecosystem and economy of the region. They vanish, and other animals and plants will vanish as well. They vanish, and the fishing industry will fold.

In this chapter, you'll see how ARC/INFO, ArcView GIS, and ArcExplorer software help conservationists restore habitat to give these fish a chance to survive and prosper.

Balancing the needs of man and nature

The U.S. Forest Service manages 156 national forests and 20 national grasslands. These 191 million acres include 128,000 miles of fishing streams and rivers, over 2.2 million acres of lakes, ponds, and reservoirs, and 12,500 miles of coast and shoreline.

These diverse lands and waters are home to more than 3,000 species of fish and wildlife and 10,000 species of plants. Of these, 283 fish, wildlife, and plant species are in need of protection under the federal Endangered Species Act.

The agency must protect these species, as well as 2,500 others designated sensitive, by balancing the needs of fishing and timber interests with the need to maintain healthy fish populations, recover threatened or endangered species, and produce fish for sport and commercial use.

Coho salmon once numbered in the hundreds of thousands and were harvested in great numbers during their annual "run," when the adults migrate upstream to deposit eggs and die.

Combining regional data

Lately, the Forest Service's mission has been complicated by the dwindling numbers of some species over a large portion of the western United States, such as cutthroat and steelhead trout and sockeye, chinook, and coho salmon.

These fish, once so abundant throughout this region, have declined during the past century, in some cases to 2 or 3 percent of former levels, over parts of Oregon, Washington, Idaho, and Utah.

Their plight has spurred federal and state agencies and Native American tribal governments to work together on restoration programs that, for the first time, take the fish's entire life cycles into account in managing their numbers and improving their chance to survive.

In 1994, and again in 1997, the agencies involved shared information about the land and streams. The stream inventory data included such information as stream width and depth, water temperature and quality, the amount of wood in the stream, and the distribution of fish. ARC/INFO software was used to enter the data into a 250-layer database covering the entire Columbia River Basin. This information was published on CD–ROM with the ArcExplorer data viewer (an entry-level GIS product) and distributed to resource managers.

Using the CD, natural resource managers can create maps like this one of dynamically segmented streams in the Pacific Northwest. Dynamic segmentation is a GIS technology that allows the user to get calculated data values for any point along a line (in this case, a stream), rather than just for those points where data was actually collected. Here, the data might include such things as estimated water quality or temperature.

Applying regional data to local projects

A Forest Service aquatics expert in Corvallis, Oregon, is using the data to develop conservation strategies for coho, a pink-and-silver salmon found mainly in the coastal river basins of Oregon and Washington.

For this study, the analyst selects the Umpqua River Basin, a system of coastal streams covering about 1 million acres. Coho were once plentiful here, but logging, home building, and "simplification," or the straightening out of rivers and streams, have damaged their habitat. The basin now has fewer than 5,000 coho, or about 3 percent of historic levels.

Salvaging coho habitat begins with identifying current coho populations and helping them expand. The analyst uses the ArcView GIS system to display and query those streams that have information about habitat and species. Several isolated reaches containing coho are found. The question now is why the fish aren't present throughout the stream network, and whether, if conditions are improved, their range can be increased.

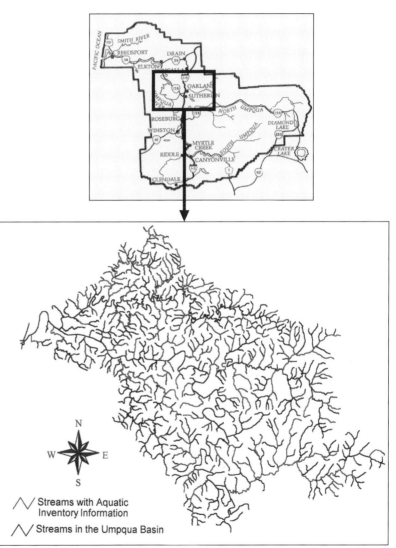

This map shows streams (in red) for which information is available. The analyst can search these stream segments for existing coho populations and identify where good habitat still exists.

Finding the right conditions

The analyst creates detailed tables and maps of the streams that could link these reaches. From here, he looks for those already suitable for coho. Three quantifiable factors indicate whether a particular stream is capable of supporting the salmon: water temperature, amount of woody debris, and existence of large pools. Each of these factors is first mapped separately.

Along with many other species, coho need streams containing large woody debris. Logs averaging 3 feet thick and over 50 feet long jam across streams, slowing the water and creating backwaters and eddies. These become vital microhabitats for the fish, providing them with places to rest, feed, and hide from predators. In addition, woody debris plays an important role in keeping stream temperatures low. Not only does it shade the water from the sun, it also causes water passing over it to scour out pools in the stream bed, where cooler groundwater is intercepted.

The analyst uses the GIS to select reaches with large pools and enough in-stream wood for coho. These are the most likely corridors for linking the existing coho populations and the ones on which he will concentrate his efforts.

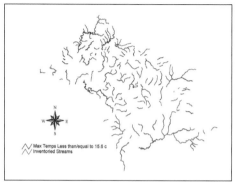

This map shows the streams (in blue) that have a low enough average temperature to support coho. The water must average 60 degrees Fahrenheit or lower.

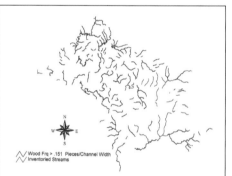

This map shows the streams (in blue) that have enough large woody debris.

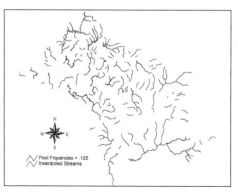

This map shows the streams (in blue) that have enough large pools for feeding and egg laying.

The final map

By combining the three layers and querying the ArcView GIS system to show the streams that meet each of the conditions, the analyst gets the map seen at the right.

The number of streams that meet all the criteria for supporting coho (shown in blue) are few. Therefore, other streams will need to be improved to support these fish. Of crucial importance is the fact that the existing coho populations are separated by segments of poor habitat. The fish are unlikely to travel these unsuitable streams and mate.

But with the streams identified, resource managers have a starting point for protecting the healthy streams and trying to restore others that will link the populations.

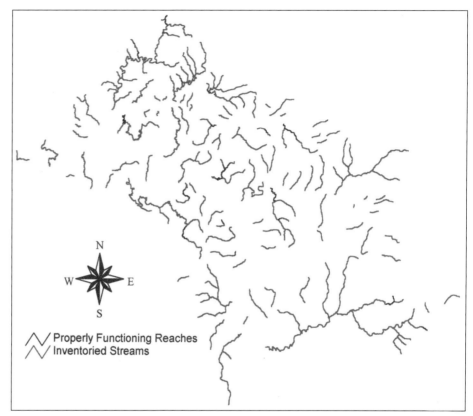

This map shows the stream reaches in the Umpqua River Basin that meet all three conditions required to support healthy populations of coho.

Habitat restoration

To link fish populations, the agency often has to coordinate with other groups—landowners, conservation groups, and fisheries experts—to improve unsatisfactory habitats. Using the CD, the analyst queries streams and ownership statistics for the entire basin to see which streams would be likely corridors. A promising spot near the top of the map, shown in dark orange, is part of the Siuslaw National Forest, managed by the Forest Service.

Since this is the case, and the streams being restored are in heavily forested and mountainous areas, the analyst suggests adding large pieces of wood to the streams. This will slow the water, cool it down, and etch holding pools, conditions that should entice the fish living nearby. By restoring these streams, the salmon in these segments will have an open path of quality habitat to each other and to the sea.

As the stream segments are restored, the agency will collect new data on the aquatic environment and fish populations and make it available to everyone participating in the project. During restoration projects, the GIS remains useful for coordinating diverse groups, creating graphics for discussion and publication, and helping agencies share new information.

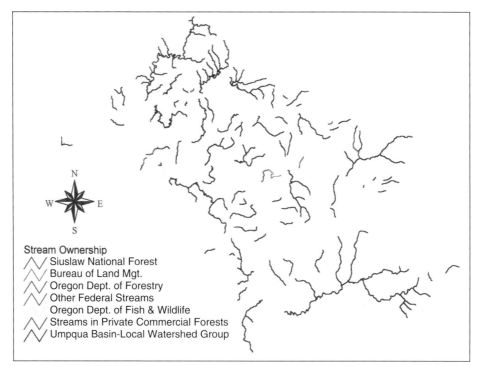

Stream Ownership
/\/ Siuslaw National Forest
/\/ Bureau of Land Mgt.
/\/ Oregon Dept. of Forestry
/\/ Other Federal Streams
    Oregon Dept. of Fish & Wildlife
/\/ Streams in Private Commercial Forests
/\/ Umpqua Basin-Local Watershed Group

This map shows which agency manages stream segments or whether the land is privately owned. Since coho travel to the ocean and back, streams must be linked all the way to the sea. This means involving private landowners, state fish and wildlife agencies, watershed councils, soil and conservation districts, and grass roots organizations.

Sustaining a healthy life cycle

In addition to the stream restoration work in river basins like the Umpqua, tribal, state, and federal agencies are working together to improve aquatic habitat along the entire migration route of the fish that live there. These projects include constructing reefs for bass and barriers for migrating salmon. The agencies are also working with conservation groups like Trout Unlimited, the Sport Fishing Institute, and the FishAmerica Foundation to set limits on fishing that will encourage repopulation of threatened and endangered species.

By 1998, project participants hope to make their data available over the Internet with several MapObjects applications. As state and federal management agencies, environmental activists, scientists, academics, lawmakers, and the public all use and contribute to the online database, the plight of the coho and other endangered salmon and trout may improve.

The agencies hope they can use the Umpqua River Basin project as a model for restoring headwaters habitat for other threatened and endangered salmon and trout.

Hardware

Sun SPARCstation workstations

IBM® 590 workstations

Pentium-class PCs

Software

ARC/INFO

ArcView GIS

ArcView Spatial Analyst

ArcExplorer

Data

Aquatic habitat and wildlife data courtesy of the Bureau of Land Management, the Oregon Department of Fish and Wildlife, and the Confederated Tribes of the Umatilla Indian Reservation.

Acknowledgments

Thanks to Shaun McKinney of Forest Service Region 6. Thanks also to the Bureau of Land Management, the Oregon Department of Fish and Wildlife, and the Confederated Tribes of the Umatilla Indian Reservation.

Chapter 12

•••• Disaster planning and recovery

Earthquakes and volcanic eruptions, while infrequent, are among nature's most devastating events. While it's impossible to predict precisely when these events will happen, it is possible to predict where they will be most destructive when they do strike. Active faults and even potential volcanic eruptions can be mapped and their impact on nearby populations and property estimated with GIS.

In this chapter, you'll see how a disaster-planning agency in Portland, Oregon, uses ARC/INFO, ArcView GIS, and MapObjects software to help local governments, businesses, emergency facilities, and residents prepare for trouble.

Understanding Portland's hazards

Portland Metro is an elected regional government serving 1.2 million people in three Oregon counties and 24 cities. Since 1993, Metro staff have spearheaded a project funded by the Federal Emergency Management Agency (FEMA) to assess how the region's buildings, roads, and other infrastructure would survive a large earthquake.

It's a common misconception that Portland—well north of the notorious San Andreas fault—isn't in "earthquake country." In reality, just off Oregon's shore lies the Cascadia subduction fault zone, stretching along the Pacific coastline from Eureka, California, to British Columbia.

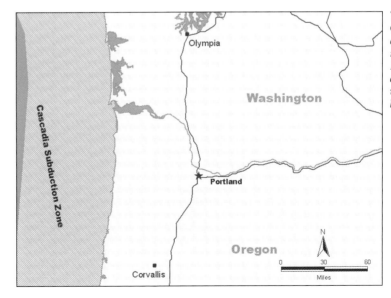

The offshore Cascadia fault concerns planners in cities like Portland, where older structures would be damaged by large quakes.

Proximity to an offshore fault

Because it's beneath the ocean, the Cascadia doesn't attract much attention. But if the fault slips, scientists warn that the earthquake would be the largest in the region's history and perhaps trigger quakes on several lesser-known but active faults.

Certain parts of Portland are especially susceptible to earthquake damage. Loosely compacted soils in areas with a high water table liquefy when the ground shakes and are unable to support buildings and roads built on top of them.

Other types of soil create different problems by amplifying ground shaking. This causes poorly constructed buildings to crumble, roads to buckle, and telephone and electrical services to be disrupted.

Following two damaging earthquakes in the early 1990s just south of Portland, the state tightened building codes. The FEMA-backed earthquake-risk project was initiated to identify which areas in the three-county region might sustain more damage from earthquakes. Identifying these areas is a first step toward making structures safer and alerting residents to the risk.

Movement on the Cascadia fault is credited with causing the 1981 eruption of Mt. St. Helens in Washington. The fault's proximity—and the area's hazardous soils—combine to make Portland a region vulnerable to building and road damage from earthquakes.

Mapping earthquake risk factors

To understand which buildings are most likely to sustain damage, the project team needed current information about soils and structures. The Oregon Department of Geology and Mineral Industries collected soil samples throughout the Portland region to see how the ground would react during a quake.

At the same time, the Portland State University Engineering Department collected information on 50,000 commercial and industrial buildings to evaluate their ability to withstand ground shaking during an earthquake.

Metro added this newly collected data to its GIS. Based on ARC/INFO software, the system has been used by the agency since 1989 to guide regional planning.

Using the GIS to overlay hazards data with information about buildings, infrastructure, and demographics, Metro's planners created earthquake hazard maps. These maps helped increase awareness of the risk in the Portland area and guided decisions to retrofit several older buildings in the region.

Many of Portland's older buildings, including its downtown city hall and main county library, were renovated to better withstand earthquakes.

Distributing risk data to planners

Metro staff decided to make its earth-quake hazard information available for distribution to regional and state emergency planners, along with GIS tools for viewing and analysis. The result was the Metropolitan Geographic Information System 1.0, built for Metro by GeoNorth, Inc., Portland, Oregon, using MapObjects and Visual Basic® software.

Metro distributed copies of this application in late 1997. The data is useful for identifying at-risk buildings. An emergency planner, for example, can use it to encourage owners and business operators to put emergency plans in place for employees and nearby residents.

She creates a map with overlays of local floodplain and soils data, highlighting the areas of greatest risk during a flood or earthquake. She then overlays buildings at these locations and uses the application's "Damage Estimation Wizard," a macro programmed by GeoNorth, to view the associated square-foot cost of repair or replacement.

The Wizard reports the total number of structures by building type. The user enters the estimated square-foot replacement cost for each to determine estimated damage costs. The application then computes structure type subtotals and generates a summary report.

The table shows the year structures were built, square footage, occupancy, use, and construction type (such as unreinforced masonry). Older buildings with unreinforced framing are more likely to sustain earthquake damage than buildings built to meet newer codes. And if these structures are located on unsafe soils, the damage could be extensive.

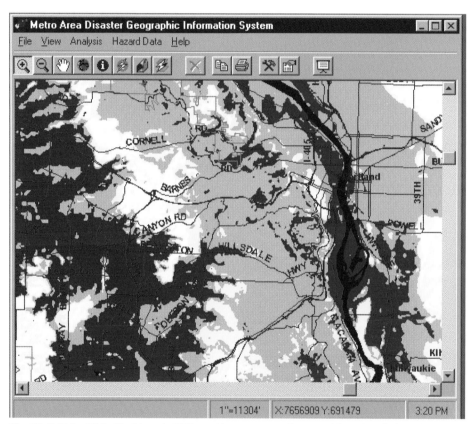

The CD–ROM published by Metro in 1997 provides emergency planners with earthquake hazard and risk data collected during the past four years. Hazardous areas are identified according to soil information and proximity to earthquake faults.

Identifying high-risk structures

The system can be queried to display buildings constructed prior to 1925 that are also located in a floodplain or built on unsafe soils. These buildings, shown in magenta, are considered high-risk structures.

With this information, the planner can meet with building owners and employers at the most populated sites to initiate earthquake preparedness programs.

Some owners might request additional help, such as neighborhood seminars or the distribution of evacuation plans to local businesses and residents. These maps can include the locations of hospitals and emergency clinics serving the community.

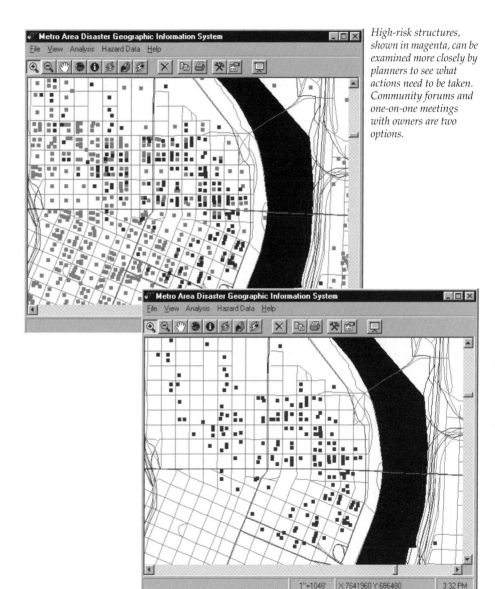

High-risk structures, shown in magenta, can be examined more closely by planners to see what actions need to be taken. Community forums and one-on-one meetings with owners are two options.

Sharing risk data with the public

Metro plans to make the CD–ROM application available to the public. It will allow nontechnical users to analyze how a quake might affect the region's vital systems such as electrical power, water, sewer, and telecommunications, as well as its probable impact on facilities like hospitals, schools, and police and fire stations.

A resident concerned about a nearby manufacturing plant can use the application to map his neighborhood and overlay flood zones and earthquake faults. He discovers that the factory is in a relatively safe zone.

He next overlays schools data and finds that his children's school is located in a floodplain and built on loosely compacted soils, making it a high-risk structure. He contacts the school's administration about evacuation planning and shares emergency route information with his wife and kids.

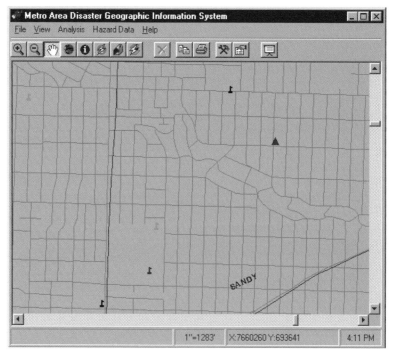

A resident uses the application to view where schools (flags) and a manufacturing plant are built in relation to sandy soil types and earthquake faults (not shown).

How businesses benefit

Meanwhile, the manufacturing company uses the application to plan what to do in a fire, flood, or earthquake, and how to get the business back up and running.

An analyst at the plant maps the surrounding community, including the locations of employee homes, fire stations, hospitals, schools, potential emergency shelters, and other employers.

As a precaution, printed evacuation maps are provided to plant employees and residents. During an emergency, this information can be shared quickly with public safety officials and sent to local media.

At a nearby hospital, the application is used by a GIS analyst to estimate how many patients to expect after a major earthquake.

The data is imported into an ArcView GIS system to create models showing differences in patient numbers depending on the time of day. If the earthquake happens during the workday, for example, the hospital can expect fewer patients than if it occurs at night when more local residents will rely on this facility.

Metro intends to make some of the earthquake risk data available over the Internet using MapObjects Internet Map Server (IMS) software. In addition to distributing the application, Metro prints risk maps and sponsors hazard-awareness workshops.

It's hoped that greater public awareness will help the region's emergency planners, businesses, residents, and emergency services prepare for earthquakes and other natural hazards.

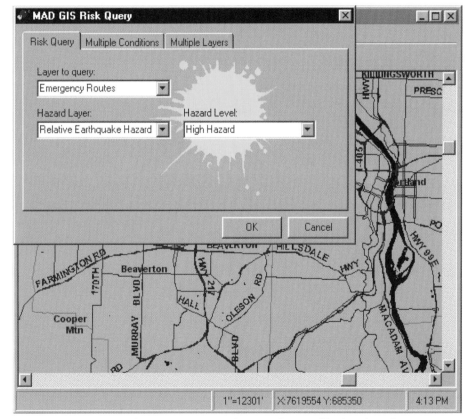

This map is used by a manufacturing company to educate employees and local residents about emergency routes to use after an earthquake.

Hardware

Hewlett–Packard® UNIX workstations

Software

ARC/INFO
ArcView GIS
MapObjects

Data

Metro's ARC/INFO database contains about 100 layers, including lot lines, tax assessor data, zoning data, comprehensive plans, parks and open space, 100-year floodplains, population, and households. This data was combined with data collected on buildings and regional seismic hazards to create the risk analysis application.

Acknowledgments

Thanks to Michael McGuire, O. Gerald Uba, and Benjamin Rice of Portland Metro, and Kellie Hauger of GeoNorth, Inc., Portland.

Portland Metro

Chapter 13

Conservation in classrooms

To help students learn about preserving the earth, many teachers have started using GIS in their classrooms. The idea is backed by universities, where these students will eventually study, and by businesses and government agencies, where they'll eventually work.

In this chapter, you'll see how students at Seaside High School in Seaside, Oregon, use ArcView GIS software to study coasts and watersheds.

The outdoor classroom

Seaside High School is a 550-student public school near the Columbia River estuary in northwestern Oregon. This location provides easy access to the Pacific Ocean, as well as its bays, inlets, tide pools, marshes, and coastal forests. For the school's Coastal Studies and Technology Center, it's the perfect setting to teach kids hands-on ecology and GIS.

Founded by biology and computer science teacher Mike Brown in 1992, the center encourages students to participate in resource-management decisions being made about Oregon's coast and wetlands.

Students work individually and in small research teams using GIS, global positioning system (GPS), video, and Internet technologies. Classroom instruction is supplemented with field time, allowing the students to work with scientists from local and federal resource management agencies, port districts, conservation groups, and universities on coastal and watershed studies. Many of their projects are featured at the school's Web site (www.seaside.k12.or.us).

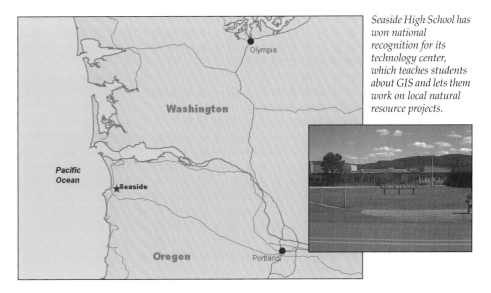

Seaside High School has won national recognition for its technology center, which teaches students about GIS and lets them work on local natural resource projects.

Collecting and publishing watershed data

Over the past few years, the center's students have collected all types of data and used technology to support projects from mapping lake bottoms to reducing speed limits on a local highway.

The soils map on this page, part of the Neawanna Watershed Education Project, was developed with the Oregon Department of Environmental Quality. The project created an online atlas (located at the school's Web site) of photographs and information about the watershed. The atlas brings together for the first time geographic, political, biological, and economic information on the watershed. The estuary data has been used by the City of Seaside and local watershed councils, saving their staffs considerable time and money.

The students started with data from the State of Oregon GIS Service Center in Salem and from the Columbia River Estuary Task Force. ArcView GIS software was used to overlay and process the data to create maps of streams, land ownership, roads, slope, zoning, soils, population density, habitat, and other information. The atlas also contains photographs taken at the site, aerial photographs, and satellite images.

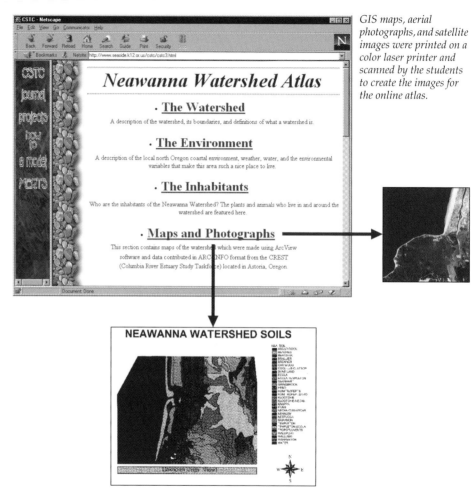

GIS maps, aerial photographs, and satellite images were printed on a color laser printer and scanned by the students to create the images for the online atlas.

Saving salmon in Trestle Bay

Another project, to restore salmon habitat in Trestle Bay on the lower Columbia River, put the center and its students in the national spotlight.

Years of sediment buildup had effectively separated Trestle Bay from the Columbia River, forming a 600-acre lagoon and raising the water's salinity. To increase circulation in the bay and improve its habitability for young salmon, the National Marine Fisheries Service was considering removing part of the bay's south jetty. About 30 students from four local high schools worked over the summer collecting samples of the bay's water. These samples were tested for temperature, brackishness, and the presence of invertebrates, such as mayflies and other insects, that young salmon eat.

The samples showed that removing the jetty would improve the bay for salmon, which prefer cold temperatures, naturally brackish water conditions, and, not too surprisingly, an abundant food supply.

The data was entered into the ArcView GIS system and overlaid with aerial photos and satellite imagery to create maps of water conditions in sampled locations. The GIS data and maps convinced the agency to remove sections of the jetty to restore habitat for fish and shorebirds. The increased

water movement and tidal scouring will increase juvenile salmon habitat and rearing areas and prevent new sediments from building up.

The center's students will return in the fall of 1998 to collect new water samples for measuring changes in the bay over the past five years.

The Trestle Bay project won the EPA's Region 10 Environmental Youth Services Award. Two students from Seaside traveled to Washington, D.C., to receive the award from President Clinton.

The CORIE Project

The students are also learning about "real-time" forecasting and computer-modeling technologies in a project for the Marine Environmental Research Training Laboratory. This $10 million research center, funded by the U.S. Navy, collects data about currents along the Columbia River estuary and makes the information available over the Internet.

The students were given a $100,000 grant to take photographs and video as a nowcast-forecast network is installed. Nowcast-forecast systems are a new forecasting technology that enables the efficient characterization of present conditions over a selected geographic area (nowcast mode) and prediction of future conditions (forecast mode). A tide gauge (an instrument that measures the time and height of tidal cycles) that is part of the system feeds information directly into the center's classroom. In the future, the students will enter data from this instrument into their ArcView GIS system to create their own maps of changing tidal conditions.

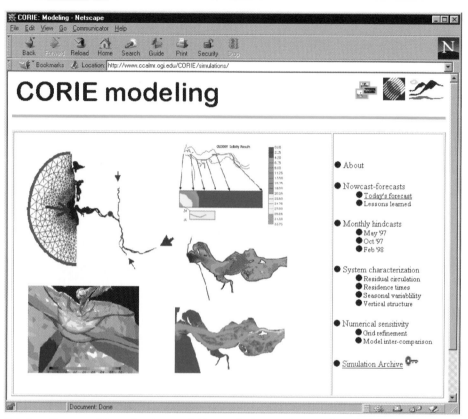

Data from the nowcast-forecast system, available over the Internet (www.ccalmr.ogi.edu), is used by state and local governments to plan how to respond to emergencies, how to navigate ships through the waters, and how to use the land—all activities that affect the lower Columbia River.

Using ArcView 3D Analyst to study runoff

Seaside's students are starting several new projects in 1998. One will study how runoff from nearby land development affects Oregon's coastal lakes. They will use a GPS echosounding instrument to record the depth and topology of lake bottoms. The data will be entered into ArcView GIS software and the ArcView 3D Analyst™ extension (an ArcView GIS add-on that displays and analyzes geographic data in three dimensions) to create 3-D maps. Many of these lake bottoms have never been mapped and the data will be the first of its kind available to local planning agencies.

Updated information will be added periodically to create a historical change database for the lake bottoms. This data will help agencies monitor silt buildups that are destroying habitat for fish, shellfish, and plants.

As shown by the projects profiled in this chapter, the GIS data being collected and stored by these students has long-term benefits for their community and the region.

The Coastal Studies and Technology Center has sparked international interest as its students share information and ideas about their projects with peers at 75 schools worldwide linked together as part of the Global Laboratory Project, sponsored by the National Science Foundation.

As the center's students move into universities and colleges, and eventually enter the workforce, they take along firsthand knowledge of how the ability to collect and analyze environmental information can make a difference in solving problems of local resource management.

Field skills, like collecting sediment and using a sieve to separate invertebrates for counting, will give these students a head start in professional life science careers.

Hardware

Pentium-based PCs

Software

ARC/INFO
ArcView GIS
ArcView 3D Analyst

Data

Aerial photography
Satellite imagery
Soils
Elevation
Water quality
Biological
Coastal
Endangered species
Political boundaries
Roads
Land use

Acknowledgments

Thanks to Mike Brown of Seaside High School and Charlie Fitzpatrick and Mike Phoenix of ESRI. The Coastal Studies and Technology Center's curriculum materials, developed with an EPA grant and assistance from ESRI and the Columbia River Estuary Study Task Force, include two workbooks and geographic data about the Columbia River. They are available for $38 from the following address:

Columbia River Estuary Task Force
750 Commercial Street, Room 214
Astoria, Oregon 97103

Telephone: 503-325-0435
Fax: 503-325-0459

**Seaside
High School**

GIS data for natural resource applications

The case studies in this book show how GIS is used by natural resource managers to make decisions. In this appendix, you'll find listed some natural resource data to get you started on your own GIS projects. Much of this data is freely available over the Internet, other data is available for purchase. You can skip immediately to these listings, or, for a quick introduction to GIS data types, read the following section on spatial and attribute data.

Spatial and attribute data

A GIS integrates two different types of data. One type defines the shape and location of places—this is called *spatial* data. The other type describes those places, the people and animals who inhabit them, and the things that happen there—this is called *attribute* data. Put succinctly, spatial data enables you to draw a map; attribute data makes the map meaningful.

Although they are structurally different, spatial and attribute data live symbiotically in a GIS, and it's hard to talk about one apart from the other. The map files you get from commercial (and noncommercial) sources normally include both.

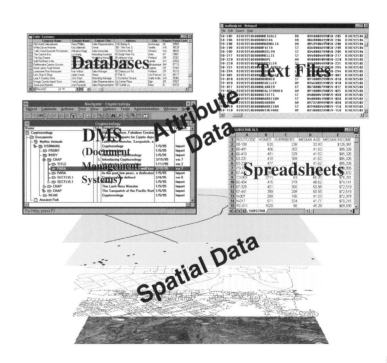

Vector, raster, and image data

Spatial data can be divided into three types. The type most common in GIS is called *vector* data. Vector data represents geographic features as points, lines, or polygons. Polygons tend to be used for sizable areas like countries or watersheds; lines for things like roads and rivers; and points to represent specific locations, like spill sites. (The shape used to represent a particular feature may depend on the scale of the map.)

Raster data takes a different approach to the mapping problem by dividing geographic space into a matrix of identically sized cells. It's commonly used to map data that's continuous—elevation, rainfall, smog levels—rather than discrete. Each cell contains a number representing the value at that location of the phenomenon being studied. The values for most cells are estimates, mathematically interpolated by the GIS from a set of actual sample measurements.

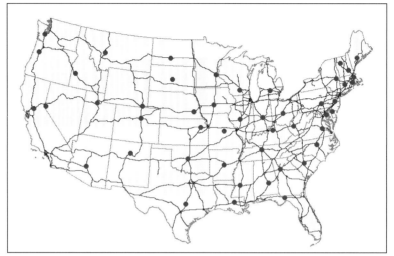

This map of the United States was created with vector data. State capitals are represented as points, interstate highways as lines, and state boundaries as polygons.

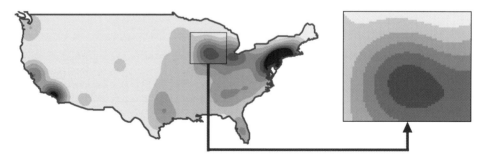

In this U.S. map, raster data is used to represent population density. Raster data is useful for depicting phenomena that are continuous and change gradually.

The third type of spatial data is *image* data, which includes such things as satellite and aerial photographs, and optically scanned paper maps. Strictly speaking, image data is a kind of raster data, because an image is composed of uniformly sized cells (pixels) linked to specific numbers—numbers that represent a color or grayness value. Image data can serve as a visual backdrop to vector data and can be used to check its accuracy. It can also be used to create vector data through a digital tracing process, or even for analysis in a process known as *remote sensing*.

ArcView GIS software uses vector spatial data (the kind of data predominantly featured in the case studies in this book); it can also display image data. ArcView Spatial Analyst software is a raster-based product that integrates seamlessly with ArcView GIS. It allows you to analyze raster data and to convert raster data to vector data.

This cloud-free satellite image of the United States from WorldSat (Burlington, Ontario) is actually a type of raster data; the data linked to each cell (pixel) represents a value of reflected light.

The latest satellites are capable of delivering 5-meter- and 1-meter-resolution images, like these of San Francisco from SpaceImaging Corp. (Thornton, Colorado).

Where to get natural resources data

Relief maps and USGS quad sheets

Chalk Butte, Inc. (Boulder, Wyoming), uses elevation data to create three-dimensional, full-color, digital relief maps. The source data is DTM, or Digital Terrain Models, from the U.S. Geological Survey (USGS). These digital images can serve as detailed topographic base maps on which to overlay vector data.

The USGS publishes detailed paper maps, known as "quad sheets," that cover the entire United States. A product called Sure!MAPS® RASTER (from Horizons Technology, San Diego, California) is comprised of scanned versions of these maps. These digital quad sheets can be used with ArcView GIS software to create new vector data or as visual backdrops.

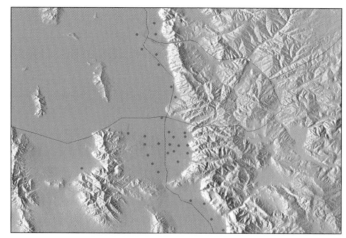

This color relief image of central Utah gives context to the vector data displayed on top of it. The lines represent major highways and the points populated places.

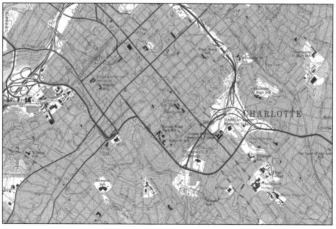

Charlotte, North Carolina, as depicted in a scanned USGS quad sheet.

Other sources of natural resources data

ArcAtlas™: Our Earth—A unique atlas of the world on CD–ROM, *ArcAtlas: Our Earth* is a large collection of maps and information about the earth—its people, plants, animals, and their various environments and economies. The atlas, produced by ESRI in conjunction with the Russian Academy of Sciences and Data+, includes more than 40 thematic maps of each continent in the form of ARC/INFO coverages and ArcView GIS projects, and more than 100 space images and photographs, plus extensive online narrative. *www.esri.com*

ArcChina Digital Map Database of China—This database consists of an overview and a map library containing 1:1,000,000-scale data. The overview, a joint effort of ESRI and the Chinese National Bureau of Surveying and Mapping, provides general aspects of the entire country, including Chinese provinces and major points of interest, and an index of the map library. The map library covers administrative boundaries down to the county level, populated places such as cities and towns, hydrography, hypsography, transportation, land cover, culture, and other natural features. *www.esri.com*

ArcScene™ World Tour—A collection of more than a dozen SPOT satellite images, jointly developed by ESRI and SPOT Image, with an educational, interactive tour in ArcView GIS software's hypertext environment. *www.esri.com*

ArcUSA™ 1:2M—ArcUSA 1:2M provides maps for the lower 48 U.S. states and counties, rivers, roads, cities, and so on. It also has 1990 and earlier demographic attributes for states and counties. *www.esri.com*

Digital Chart of the World—The largest-scale (1:1,000,000) database covering the earth's land area uniformly. It's available for use with ARC/INFO and ArcView GIS software. *www.esri.com*

Iowa DNR - Natural Resource GIS Library—This site contains the Iowa Department of Natural Resources GIS library of PC ARC/INFO® coverages. Export files are tarred and compressed. *www.igsb.uiowa.edu*

National Wetlands Inventory—The National Wetlands Inventory produces maps at a scale of 1:24,000. These maps contain information on the characteristics, extent, and status of the nation's wetlands and deep-water habitats. The maps are digitized and put on the Web in ARC/INFO export and DLG formats. *www.nwi.fws.gov*

Wyoming Natural Resources Data Clearinghouse—A source of spatial data developed by the University of Wyoming and affiliated organizations. The primary focus is on natural resources data such as geology, hydrology, biodiversity, water rights, and political jurisdictions. Also includes some industry-related data. *www.sdvc.uwyo.edu*

Verde River Watershed—This site contains multiple coverages for the extent of the Verde River watershed in central Arizona. *www.verde.org/covers.htm*

Gifford Pinchot National Forest—A source of spatial and tabular data sets for the Gifford Pinchot National Forest, which covers approximately 1.4 million acres in the state of Washington. Spatial data is provided in ARC/INFO export format and tabular data in ASCII delimited format. *www.fs.fed.us/gpnf*

InfoRain—InfoRain is the bioregional information system for the North American coastal rain forest, providing accessible and useful information at multiple scales. *www.inforain.org*

VENTS—NOAA's VENTS program, based in Newport, Oregon, and Seattle, Washington, studies hydrothermal venting systems primarily in the northeast Pacific Ocean. Research includes bathymetry mapping, submersible operations, earthquake monitoring, water column sampling, underwater photography, and other sampling. The VENTS data page provides access to currently available downloadable data sets as well as map exploration using Internet mapping technology. *newport.pmel.noaa*

Environmental Protection Agency— The EPA's node on the National Geospatial Data Clearinghouse, a component of the National Spatial Data Infrastructure (NSDI). The node provides a pathway to geospatial data at the EPA. *www.epa.gov*

University of Rhode Island— Comprehensive GIS data (raster and vector) for Rhode Island. Data sets may either be statewide or by 1:24,000 USGS quadrangle. *edc.uri.edu*

State of Kansas Data Access and Support Center—The State of Kansas GIS Policy Board's Data Access and Support Center (DASC) was established in 1991 to administer access to the state's growing GIS database. Much of the Kansas GIS core database is available free of charge from the DASC Web site. *gisdasc.kgs.ukans.edu*

Socioeconomic Data and Applications Center—This site, which provides access to socioeconomic data and information at many institutions and data centers around the world, is vital to interdisciplinary assessment of global change issues. *wwwgateway.ciesin.org*

Africa Land Cover—One portion of a global land cover characteristics database being developed for all continents. Also includes a core set of derived thematic maps produced through the aggregation of seasonal land cover regions. *edcwww.cr.usgs.gov/landdaac/glcc/at_int.html*

Environmental Indicators—The Annotated Digital Atlas of Global Water Quality selectively summarizes data contributed by countries participating in the GEMS/WATER Programme from 1976 to 1990. The site also includes information from other sources and summarizes the results of analysis and interpretation for 82 major river basins around the world. Some of these are located in highly populated and industrialized areas where human impact can be clearly seen. Others represent more pristine areas that are currently not under extensive human stress. *www.epa.gov/indicators/index.html*

The Earth Resources Observation Systems (EROS) Data Center Distributed Active Archive Center (EDC DAAC)— Established as part of NASA's Earth Observing System Data and Information System (EOSDIS) initiative to promote the interdisciplinary study and understanding of the integrated earth system. This site provides access to land processes data, including satellite- and aircraft-acquired data stored in the EDC DAAC's archives. *edcwww.cr.usgs.gov/landdaac*

South American Digital Elevation Model (DEM)—This 30-arc-second digital elevation model of South America was developed from five data sources: DTED (Digital Terrain Elevation Data, Defense Mapping Agency); DCW (Digital Chart of the World, Defense Mapping Agency); IMW (International Map of the World); AMS (Army Map Service); and PERU (map of Peru, Government of Peru). *edcwww.cr.usgs.gov/landdaac/gtopo30/ papers/olsen.html*

USGS Earth Science Data—Includes cartographic data, climate data, geologic data, hydrologic data, land use/land cover, satellite and aerial photograph data, soils data, photographs, and printed maps. *edcwww.cr.usgs.gov/webglis*

World Conservation Monitoring Centre—Databases on the world's threatened animals and plants, including access to the 1996 IUCN Red List of Threatened Animals and the new 1997 IUCN Red List of Threatened Plants. *www.wcmc.org.uk*

Coral Reef Bleaching Events—These events have been noted in areas where the sea surface temperature (SST) exceeds the climatological maximum for that region by 1 degree centigrade or more. *psbsgi1.nesdis.noaa.gov:8080/PSB/EPS/SST/climohot.html*

Gridded Population of the World—Demographic information by ecological zones (as recorded with satellite imagery) rather than divided by nation states. *www.ciesin.org/datasets/gpw/globldem.doc.html*

Glossary

ARC/INFO   A GIS software package from ESRI that runs on UNIX workstations and Windows NT. Map files created with ARC/INFO software may be used in ArcView GIS projects.

ArcView Internet Map Server   An ArcView GIS extension that supports live mapping and GIS applications on the World Wide Web.

ArcView Spatial Analyst   An ArcView GIS extension that supports spatial and statistical analysis of raster data and the integrated use of raster and vector data.

attribute   A piece of information describing a map feature. The attributes of a ZIP Code, for example, might include its area, population, and average per capita income. Attribute data is one of the two main types of data in a GIS (the other being spatial data).

biodiversity   Indication of the variety of species native to a particular habitat. When some species are reduced in numbers, biodiversity is said to be threatened.

biomass   The amount of living matter (i.e., plants and animals) in a given area.

brownfields   Vacant or underutilized land parcels with soils that may be contaminated.

browser   Client software used to access resources on the World Wide Web.

corridors   Land or streams that link isolated groups of species for breeding and migratory purposes.

dBASE file	A file format native to dBASE data management software. ArcView GIS software can read, create, and export tables in dBASE format.
demographics	The statistical characteristics of a population (for example, income, education, and home ownership).
digital elevation model	A topographic surface arranged in a data file as a set of x,y,z coordinates where z equals elevation.
digitizing	The process of electronically tracing features on a paper map to convert them to features in a digital map file. Accomplished with a device called a digitizing tablet.
Economic Development Zone	A region within a city that may qualify for state-funded projects or tax incentives to attract business and development.
ecosystem	A community of organisms and its environment, functioning as an ecological unit in nature.
Empowerment Zone	A designated area in which businesses can qualify for federal financial support.
fault zone	An area underground where cracks in the earth indicate tectonic plate movements. These are areas where earthquakes are likely to occur.
feature	A map representation of a geographic object. Land parcels, streams, species habitats, and pollution pour points are examples of map features. Features are drawn in ArcView GIS software as points, lines, and polygons.
floodplain	Level land that may be submerged by floodwaters.
geographic information system (GIS)	A configuration of computer hardware and software that stores, displays, and analyzes geographic data.

growth zone	Area where existing or planned infrastructure would support more homes or businesses. These areas are targeted by cities for development to steer growth away from open space, agricultural land, wetlands, and other nondevelopable areas.
Global Positioning System (GPS)	The GPS consists of 24 satellites that send signals to receivers on the ground. The signals provide latitude/longitude locations that can be entered into a GIS.
habitat	The place where an animal or plant normally lives.
hazard data	Information used to gauge risks to structures and lives, such as earthquake fault data, floodplain data, and information about unsafe buildings.
image data	One of the three types of spatial data in a GIS (the others being raster and vector data). Photographs taken from satellites and airplanes are examples of image data.
Internet	A decentralized computer network linking tens of thousands of smaller networks and accessed by more than 30 million users worldwide. Users connected to the Internet can send and receive e-mail, download files, view multimedia content on the World Wide Web, and run software applications stored on remote computers.
intranet	A computer network with restricted access (as, for example, within a company) that uses standard Internet protocols like HTML and HTTP.
land recycling	The redevelopment of vacant or underutilized urban parcels.
layer	A set of related map features and attributes, stored as a unique file in a geographic database. A GIS can display multiple layers (for instance, buildings, fault lines, and soil types) at the same time.
MapObjects	An ESRI GIS product that enables developers to publish interactive maps on Web sites.

pH A scale from 0–14 for expressing acidity and alkalinity.

pollution point source A place, such as a factory or farm, where toxic wastes are produced.

raster data One of the three types of spatial data in a GIS (the others being image and vector data). Raster data represents geographic space as a matrix of cells; map features are defined by numeric values assigned to the cells.

remote sensing The acquisition of data from a distance, such as by satellite imagery or aerial photography.

runoff Water from rain or snow that travels downstream, often carrying soil and wastes.

shapefile A file format developed by ESRI for storing the location, shape, and attribute information of geographic features.

spatial analysis The determination of the spatial relationships between geographic objects, such as the distance between them or the extent to which they overlap.

spatial data One of the two main types of data in a GIS (the other being attribute data). Spatial data represents the shape, location, or appearance of geographic objects. It can be in vector, raster, or image format.

symbol A particular graphic element or icon (defined by some combination of shape, size, color, angle, outline, and fill pattern) used to draw a map feature. An airport, for example, might be represented by an icon of a blue airplane. ArcView GIS software comes with hundreds of symbols to choose from; additional symbols can be created from fonts or imported from images.

thematic map A map that symbolizes features according to a particular attribute. A map displaying stream segments in different colors according to their degree of pollution is an example.

watershed An area where streams and rivers drain into a common body of water.

URL Uniform Resource Locator. The address for a site on the World Wide Web, like http://www.esri.com.

vector data One of the three types of spatial data in a GIS (the others being image and raster data). Vector data represents geographic objects as points, lines, or polygons.

World Wide Web A client/server system for distributing and accessing multimedia documents on the Internet. Documents on the World Wide Web are formatted in a special language called HTML (HyperText Markup Language) that supports links to other documents.

Other books from **ESRI Press**

GIScience

The ESRI Guide to GIS Analysis, Volume 1: Geographic Patterns and Relationships
An important book about how to do real analysis with a geographic information system. *The ESRI Guide to GIS Analysis* focuses on six of the most common geographic analysis tasks. ISBN 1-879102-06-4 188 pages

Modeling Our World: The ESRI Guide to Geodatabase Design
With this comprehensive guide and reference to GIS data modeling and to the new geodatabase model introduced with ArcInfo™ 8, you'll learn how to make the right decisions about modeling data, from database design and data capture to spatial analysis and visual presentation.
ISBN 1-879102-62-5 216 pages

Hydrologic and Hydraulic Modeling Support with Geographic Information Systems
This book presents the invited papers in water resources at the 1999 ESRI® International User Conference. Covering practical issues related to hydrologic and hydraulic water quantity modeling support using GIS, the concepts and techniques apply to any hydrologic and hydraulic model requiring spatial data or spatial visualization. ISBN 1-879102-80-3 232 pages

Beyond Maps: GIS and Decision Making in Local Government
Beyond Maps shows how local governments are making geographic information systems true management tools. Packed with real-life examples, it explores innovative ways to use GIS to improve local government operations. ISBN 1-879102-79-X 240 pages

The ESRI Press Dictionary of GIS Terminology
This long-needed and authoritative reference brings together the language and nomenclature of the many GIS-related disciplines and applications. Designed for students, professionals, researchers, and technicians, the dictionary provides succinct and accurate definitions of more than a thousand terms. ISBN 1-879102-78-1 128 pages

Planning Support Systems: Integrating Geographic Information Systems, Models, and Visualization Tools
Richard Brail of Rutgers University's Edward J. Bloustein School of Planning and Public Policy, and Richard Klosterman of the University of Akron, have assembled papers from colleagues around the globe who are working to expand the applicability and understanding of the top issues in computer-aided planning. ISBN 1-58948-011-2 468 pages

Geographic Information Systems and Science
This comprehensive guide to GIS, geographic information science (GIScience), and GIS management illuminates some shared concerns of business, government, and science. It looks at how issues of management, ethics, risk, and technology intersect, and at how GIS provides a gateway to problem solving, and links to special learning modules at ESRI Virtual Campus (campus.esri.com). ISBN 0-471-89275-0 472 pages

Undersea with GIS
Explore how GIS is illuminating the mysteries hidden in the earth's oceans. Applications include managing protected underwater sanctuaries, tracking whale migration, and recent advances in 3-D electronic navigation. The companion CD brings the underwater world to life for both the undersea practitioner and student and includes 3-D underwater flythroughs, ArcView® extensions for marine applications, a K–12 lesson plan, and more.
ISBN 1-58948-016-3 276 pages

Past Time, Past Place: GIS for History
In this pioneering book that encompasses the Greek and Roman eras, the Salem witch trials, the Dust Bowl of the 1930s, and much more, leading scholars explain how GIS technology can illuminate the study of history. Richly illustrated, *Past Time, Past Place* is a vivid supplement to many courses in cultural studies and will fascinate armchair historians.
ISBN 1-58948-032-5 224 pages

Arc Hydro: GIS for Water Resources
Based on ESRI ArcGIS™ software, the Arc Hydro data model provides a new and standardized way of describing hydrologic data to consistently and efficiently solve water resource problems at any spatial scale. This book is a blueprint of the model and the definitive overview of GIS in hydrology from the field's leading expert. The companion CD includes Arc Hydro instructions, tools, teacher resources, and data. ISBN 1-58948-034-1 220 pages

Confronting Catastrophe: A GIS Handbook
This hands-on manual is for GIS practitioners and decision makers whose communities face the threat of large-scale disasters, whether they be wildfires, hurricanes, earthquakes, or now, terrorist attacks. Real-world lessons show public officials and IT managers how to use GIS most efficiently in the five stages of disaster management: identification and planning, mitigation, preparedness, response, and recovery. ISBN 1-58948-040-6 156 pages

A System for Survival: GIS and Sustainable Development
The goal of sustainable development is to raise standards of living worldwide without depleting resources or destroying habitat. This book provides examples of how geographic technologies, by synthesizing the vast amounts of data being collected about natural resources, population, health, public safety, and more, can play a creative and constructive role in the realization of that goal. ISBN 1-58948-052-X 124 pages

Marine Geography: GIS for the Oceans and Seas
This collection of case studies documents some of the many current applications of marine GIS. The contributing authors share their work, challenges, successes, and progress in the use of GIS, and show how the technology is being used to influence the decision-making process in a way that leads to healthy and sustainable oceans and seas.
ISBN 1-58948-045-7 228 pages

CONTINUED ON NEXT PAGE

Other books from **ESRI Press** *continued*

The Case Studies Series

ArcView GIS Means Business

Written for business professionals, this book is a behind-the-scenes look at how some of America's most successful companies have used desktop GIS technology. The book is loaded with full-color illustrations and comes with a trial copy of ArcView software and a GIS tutorial. ISBN 1-879102-51-X 136 pages

Zeroing In: Geographic Information Systems at Work in the Community

In twelve "tales from the digital map age," this book shows how people use GIS in their daily jobs. An accessible and engaging introduction to GIS for anyone who deals with geographic information. ISBN 1-879102-50-1 128 pages

Serving Maps on the Internet

Take an insider's look at how today's forward-thinking organizations distribute map-based information via the Internet. Case studies cover a range of applications for ArcView Internet Map Server technology from ESRI. This book should interest anyone who wants to publish geospatial data on the Web. ISBN 1-879102-52-8 144 pages

Managing Natural Resources with GIS

Find out how GIS technology helps people design solutions to such pressing challenges as wildfires, urban blight, air and water degradation, species endangerment, disaster mitigation, coastline erosion, and public education. The experiences of public and private organizations provide real-world examples. ISBN 1-879102-53-6 132 pages

Enterprise GIS for Energy Companies

A volume of case studies showing how electric and gas utilities use geographic information systems to manage their facilities more cost effectively, find new market opportunities, and better serve their customers. ISBN 1-879102-48-X 120 pages

Transportation GIS

From monitoring rail systems and airplane noise levels, to making bus routes more efficient and improving roads, the twelve case studies in this book show how geographic information systems have emerged as the tool of choice for transportation planners. ISBN 1-879102-47-1 132 pages

GIS for Landscape Architects

From Karen Hanna, noted landscape architect and GIS pioneer, comes *GIS for Landscape Architects*. Through actual examples, you will learn how landscape architects, land planners, and designers now rely on GIS to create visual frameworks within which spatial data and information are gathered, interpreted, manipulated, and shared. ISBN 1-879102-64-1 120 pages

GIS for Health Organizations

Health management is a rapidly developing field, where even slight shifts in policy affect the health care we receive. In this book, you will see how physicians, public health officials, insurance providers, hospitals, epidemiologists, researchers, and HMO executives use GIS to focus resources to meet the needs of those in their care. ISBN 1-879102-65-X 112 pages

GIS in Public Policy: Using Geographic Information for More Effective Government

This book shows how policy makers and others on the front lines of public service are putting GIS to work—to carry out the will of voters and legislators, and to inform and influence their decisions. *GIS in Public Policy* shows vividly the very real benefits of this new digital tool for anyone with an interest in, or influence over, the ways our institutions shape our lives. ISBN 1-879102-66-8 120 pages

Integrating GIS and the Global Positioning System

The Global Positioning System is an explosively growing technology. *Integrating GIS and the Global Positioning System* covers the basics of GPS and presents several case studies that illustrate some of the ways the power of GPS is being harnessed to GIS, ensuring, among other benefits, increased accuracy in measurement and completeness of coverage. ISBN 1-879102-81-1 112 pages

GIS in Schools

GIS is transforming classrooms—and learning—in elementary, middle, and high schools across North America. *GIS in Schools* documents what happens when students are exposed to GIS. The book gives teachers practical ideas about how to implement GIS in the classroom, and some theory behind the success stories. ISBN 1-879102-85-4 128 pages

Disaster Response: GIS for Public Safety

GIS is making emergency management faster and more accurate in responding to natural disasters, providing a comprehensive and effective system of preparedness, mitigation, response, and recovery. Case studies describe GIS use in siting fire stations, routing emergency response vehicles, controlling wildfires, assisting earthquake victims, improving public disaster preparedness, and much more. ISBN 1-879102-88-9 124 pages

Open Access: GIS in e-Government

A revolution taking place on the Web is transforming the traditional relationship between government and citizens. At the forefront of this e-government revolution are agencies using GIS to serve interactive maps over their Web sites and, in the process, empower citizens. This book presents case studies of a cross-section of these forward-thinking agencies. ISBN 1-879102-87-0 124 pages

GIS in Telecommunications

Global competition is forcing telecommunications companies to stretch their boundaries as never before—requiring efficiency and innovation in every aspect of the enterprise if they are to survive, prosper, and come out on top. The ten case studies in this book detail how telecommunications competitors worldwide are turning to GIS to give them the edge they need. ISBN 1-879102-86-2 120 pages

Conservation Geography: Case Studies in GIS, Computer Mapping, and Activism

This collection of dozens of case studies tells of the ways GIS is revolutionizing the work of nonprofit organizations and conservation groups worldwide as they rush to save the earth's plants, animals, and cultural and natural resources. As these pages show clearly, the power of computers and GIS is transforming the way environmental problems and conservation issues are identified, measured, and ultimately, resolved. ISBN 1-58948-024-4 252 pages

GIS Means Business, Volume Two

For both business professionals and general readers, *GIS Means Business, Volume Two* presents more companies and organizations, including a chamber of commerce, a credit union, colleges, reinsurance and real estate firms, and more, who have used ESRI software to become more successful. See how businesses use GIS to solve problems, make smarter decisions, enhance customer service, and discover new markets and profit opportunities. ISBN 1-58948-033-3 188 pages

ESRI Software Workbooks

Understanding GIS: The ARC/INFO® Method (Version 7.2 for UNIX® and Windows NT®)

A hands-on introduction to GIS technology. Designed primarily for beginners, this classic text guides readers through a complete GIS project in ten easy-to-follow lessons. ISBN 1-879102-01-3 608 pages

ARC Macro Language: Developing ARC/INFO Menus and Macros with AML

ARC Macro Language (AML™) software gives you the power to tailor workstation ARC/INFO software's geoprocessing operations to specific applications. This workbook teaches AML in the context of accomplishing practical workstation ARC/INFO tasks, and presents both basic and advanced techniques. ISBN 1-879102-18-8 828 pages

Getting to Know ArcView GIS

A colorful, nontechnical introduction to GIS technology and ArcView software, this best-selling workbook comes with a working ArcView demonstration copy. Follow the book's scenario-based exercises or work through them using the CD and learn how to do your own ArcView project. ISBN 1-879102-46-3 660 pages

Extending ArcView GIS

This sequel to the award-winning *Getting to Know ArcView GIS* is written for those who understand basic GIS concepts and are ready to extend the analytical power of the core ArcView software. The book consists of short conceptual overviews followed by detailed exercises framed in the context of real problems. ISBN 1-879102-05-6 540 pages

GIS for Everyone, Second Edition

Now everyone can create smart maps for school, work, home, or community action using a personal computer. This revised second edition includes the ArcExplorer™ geographic data viewer and more than 500 megabytes of geographic data. ISBN 1-879102-91-9 192 pages

Getting to Know ArcGIS Desktop: Basics of ArcView, ArcEditor™, and ArcInfo

Getting to Know ArcGIS Desktop is a workbook for learning ArcGIS, the newest GIS technology from ESRI. Learn to use the building blocks of ArcGIS: ArcMap™, for displaying and querying maps; ArcCatalog™, for managing geographic data; and ArcToolbox™, for setting map projections and converting data. Illustrations and exercises teach basic GIS tasks. Includes a 180-day trial version of ArcView 8 software on CD and a CD of data for working through the exercises. ISBN 1-879102-89-7 552 pages

Mapping Our World: GIS Lessons for Educators

A comprehensive educational resource that gives any teacher all the tools needed to begin teaching GIS technology in the middle- or high-school classroom. Includes nineteen complete GIS lesson plans, a one-year license of ArcView 3.x, geographic data, a teacher resource CD, and a companion Web site. ISBN 1-58948-022-8 564 pages

CONTINUED ON NEXT PAGE

Atlases

Mapping Census 2000: The Geography of U.S. Diversity
Cartographers Cynthia A. Brewer and Trudy A. Suchan have taken Census 2000 data and assembled an atlas of maps that illustrates the new American diversity in rich and vivid detail. The result is an atlas of America and of Americans that is notable both for its comprehensiveness and for its precision. ISBN 1-58948-014-7 120 pages

Salton Sea Atlas
Explore the scientific, historical, and physical dimensions of California's unique Salton Sea region, a nexus of the forces of nature and the momentum of human civilization, in the pages of this groundbreaking benchmark reference. Cutting-edge GIS technology was used to transform vast amounts of data into multilayered, multithemed maps as visually compelling as they are revelatory. ISBN 1-58948-043-0 136 pages

ESRI Map Book, Volume 17: Geography and GIS—Sustaining Our World
A full-color collection of some of the finest maps produced using GIS software. Published annually since 1984, this unique book celebrates the mapping achievements of GIS professionals. *Directions Magazine* (www.directionsmag.com) has called the *ESRI Map Book* "The best map book in print." ISBN 1-58948-048-1 120 pages

ESRI educational products cover topics related to geographic information science, GIS applications, and ESRI technology. You can choose among instructor-led courses, Web-based courses, and self-study workbooks to find education solutions that fit your learning style and pocketbook. Visit www.esri.com/education for more information.

ESRI Press publishes a growing list of GIS-related books. Ask for these books at your local bookstore or order by calling 1-800-447-9778. You can also shop online at www.esri.com/gisstore. Outside the United States, contact your local ESRI distributor.